NUMBER 225

THE ENGLISH EXPERIENCE

ITS RECORD IN EARLY PRINTED BOOKS
PUBLISHED IN FACSIMILE

JOHN BLAGRAVE

BACULUM FAMILIARE

LONDON 1590

DA CAPO PRESS
THEATRVM ORBIS TERRARVM LTD.
AMSTERDAM 1970 NEW YORK

The publishers acknowledge their gratitude
to the Curators of the Bodleian Library, Oxford
for their permission to reproduce
the Library's copy (Shelfmark: Vet. A1.e.51)
and to the Syndics of Cambridge University Library
for their permission to reproduce
the leaves: D2 verso, D3 recto, D4 verso. K3 verso
from their copy (Shelfmark: Syn.5.59.16).

S.T.C. No. 3118
Collation: A-K^4

Published in 1970 by
Theatrum Orbis Terrarum Ltd.,
O.Z. Voorburgwal 85, Amsterdam

&

Da Capo Press
- a division of Plenum Publishing Corporation -
227 West 17th Street, New York, 10011
Printed in the Netherlands

ISBN 90 221 0225 4

Baculum Familliare, Catholicon
siue Generale.

A BOOKE OF THE
making and vse of a Staffe, newly inuented by the Author, called the

Familiar Staffe.

As well for that it may be made vsually and familiarlie to walke with, as for that it performeth the Geometrical mensurations of all Altitudes, Longitudes, Latitudes, Distances and Profundities: as many myles of, as the eye may well see and discerne: most speedily, exactly and familiarly without any maner of Arithmeticall calculation, easily to be learned and practised, euen by the vnlettered.

Newlie compiled, and at this time published for the speciall helpe of shooting in great Ordinance, and other millitarie seruices, and may as well be imployed by the ingenious, for measuring of land, and to a number of other good purposes, both Geometricall and Astronomicall:

By
IOHN BLAGRAVE of Reading Gentleman,
the same well willer to the Mathematickes.

The Vse of which Familiar staffe is also so generall that it readily performeth all the seuerall Vses of the Crosse staffe, *the* Quadrate, *the* Circle, *the* Quadrante, *the* Gunners Quadrante, *the* Trigon, *every one in his owne kinde, and with no lesse methode and facillitie, both for Sea and Land.*

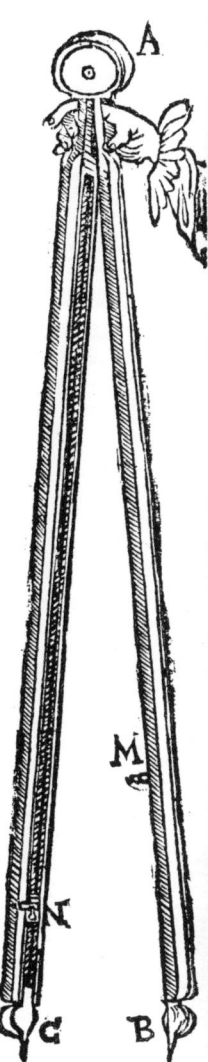

LONDON
Printed by Hugh Iackson dwelling in Fleetestreete a litle beneath the Conduit, at the signe of S. Iohn the Euangelist.
1590.

To the Right Honourable Sir Fraunces Knolles Knight, Treasurer of her Maiesties Houshold, and of her Highnesse most Honourable priuie Counsell.

When that I had (right Honourable) with an earnest intent to proceede very farre in the Mathematikes published my late booke of the *Mathematicall Iewell*, dedicated euen of meere duety, vnto the right Honourable Lord Burleigh, Lord high treasurer of England, whom from the aboundance of my hart I can not but with your Honours good fauour, as a Myrror of Iustice and pietie, a Patterne of true Honour and Nobilitie, not onely at this time againe, but during life, in all my good actions to remember. Yet not long after espying my yeares to run on, the world to slip away by me, whiles I was occupied in these studies that brought nothing but delight: and withall remembring the example of the silly Grashopper, and the saying *non semper erit estas*. I began in maner to droupe and languish, as one out of hope, and euen vpon point either to giue them ouer, or to attend them at more leasure, which of all thinges those studies may not like because they rather more earnestly require a mans whole indeuour, Sticking fast in this mammering of dispaire, your Honour beyond all exspectation or cogitation of mine, drawes me forth of the poore countrie Cabin where obscurely I lurked, into your more open presence, and there so adorned and beautified me with your Honourable curtesies and fauour, so renoumed my name at the Court amongst persons of high regard, and (not content therewith) without any least desart of mine, of purpose, as I take it, to encourage me to proceed: thereby not a litle bewraying your Honours hidden skill and secret good will to the Mathematicke sciences, most liberally bestowed on me a yearely pension or stipend: that verifying the old Adage: *Honos alit artes*, I was enforced againe to my former Byas, and so much the more strongly, to the end that your Honour conceaued opinion of me, and fauourable auouchments in my behalfe, should not be vtterly made void, earnestly bending vpon those pointes which may most helpe or forward Nauigators for their long voyages and new discoueries being now almost prepared and in point, to furnish them with such familiar instruments and precepts, carrying no lesse facilitie then this treatise doth, that them selues shall bee able to strengthen their owne nauigations, in whatsoeuer arte may assist them: But lest

The Epistle Dedicatorie.

in the meane time whiles I am thus busied, euen to the vttermost of that leasure which this diuers world alloweth me, I should be thought to sleepe of your Honour, of whom I haue beene so many waies fauourably incyted, I could not but exact so much intermission from those waightie actions, as to seeke about in my Mathematicke storehouse, for some ready present wherewith I might the whiles shew my selfe broad waking vnto your H. And all sodeinly as I was tossing and seeking, me thought I hard the fresh sound in mine eares of the peece of Ordinance I saw your H. shoote off at Greyes this last sommer. Thereupon I sought no further, but seeing your H. both by your honourable office and course of yeares to carry a staffe, with that conceite bent my head wholy to metamorphose that instrument which by chaunce I then shewed vnto your H. & others there: vnto a Mathematicall staffe, fit for so noble a mind as your H. hath, euen frō your first knighthood, wonne in the field, hetherto carried, and dayly expresseth, by diuers such noble and martiall exercises. And the same staffe cleane void of all loftie florishes, singuler easie to be conceiued, euen for the vnlettered, to learne and worke by, without any manner of arithmeticall calculation, although I know your H. most aboundantly farre to exceede any ordinary Arithmetician, and not deuoid of some partes Geometricall. Which staffe, as effectually framed as my selfe and my man (whom purposly to forward these actions, I haue euer since your Honours saide liberalitie bestowed, retained) cold without any first patterne contriue: I most humbly present to your H. together with this treatise of the vse thereof, beseeching God to prosper your H. in all your honourable actions.

Your Honours most humble at command,

Iohn Blagraue.

A Table of the Chapters and Contentes of this present booke following.

Chap. 1. What moued the Author at this time to publish so much of this instrument, & his vse as he now setteth foorth.

Chap. 2. Of the imperfections of the crosse staffe, called by *Gemma Frisius Baculus Geometricus* and by howe much this Familliar Staffe exceedeth it and all other instrumentes hetherto deuised, to like purposes both for sufficiency and facillity.

Chap. 3. Of the framing and fashioning of this notable instrument called the Familiar Staffe.

Chap. 4. How you shall in a singular sort set degrees on the leuel of the running staffe, and also on his Graduator, together with the pointes of the Gunners quadrant.

Chap. 5. How by this Familiar Staffe to leuell or try, whether a peece of ground be leuell whereon to plant your peece of ordinance.

Chap. 6. How by this Familliar Staffe to mount a piece of ordinance by points of the Gunners quadrant.

Chap. 7. By what meanes a certaine Table is to be made, therby to know how farre any piece wil shoot at random, being mounted to any point of the gunners quadrant.

Chap. 8. Hauing made a Table of Randomes to some one piece according to the precept in the last chapter. How by this Familiar Staffe, to make the same table serue for any other piece, without any Arythmeticall calculation singular easie.

Chap. 9. If a Wall or Tower were to be scaled, and that you may come vnto the base of it without daunger, How by this Familiar staffe speedely to get the height thereof, thereby to make your scaling ladders accordjng.

Chap. 10. To performe the last chap. Where you dare not come neere the base of the Tower for daunger of shot, or let by reason of some deepe mote or ditch.

Chap. 11. How by this Familiar staffe to performe the last chap. another way more exact for long distances, the more safely to keepe you out of daunger of shot of the Fort, whiles you are in action.

Chap. 12. How you shall knowe by this Familiar staffe, the depth vnder the iust leuell of your eye, of the base of any Tower vnapprochable, when the same base is to be seene.

Chap. 13. To know the length of the scaling ladder to reach ouer the ditch to the toppe of the wall or tower.

Chap. 14. How at any station, eyther by the standerd or running staffe, the angle of station or position betweene any two markes or places, is to be taken two seuerall wayes.

Chap. 15. How (in manner of the first and playnest meanes mentioned in the last chapter, to take the angle of position) to get the most

The Table.

most exact distance of a castle or fort from you, though the same fort be two or three myles off or more, whereby you may know how to place your mayne battell, as neere as may be, yet without danger of shot from the Fort, and also in what space you may march to the same, when you will.

Chap. 16. Of the Geometricall ground and familiar proofe thereof, whereon the whole working by this familiar Staffe dependeth.

Chap. 17. How to performe the fifteenth chapter with more facility by meanes of the second manner of taking the angle of position mentioned in the 14. chap.

Chap. 18. In case that a Castle or Fort were digged out of some rocke or scituate in some valley betweene two hilles, admitting but some narrowe comming vnto it, and therefore enforcing your stations to be one directly behinde another, How yet by this Familiar Staffe you shall attaine the distance of the same from you.

Chap. 19. How you shall performe the 15. chap. Where the distance of the fort or castle is very farre off, & the ground being vneuen with hilles, dales, and rockes, hath no one leuell plaine sufficient to make two stations for so great a distance.

Chap. 20. If in a night you haue secretly gotten with your army neere any Fort, and that you would with more speede then in the 15. or 17. chap. is shewed, knowe the distance whether you are neere ynough to plant your ordinance for battery.

Chap. 21. If you shall see two Fortes of the Enemies within biewe, and would knowe how farre they are in sunder, and whether there may be passage for an Army betweene them without daunger of shot from those Fortes, or to get the length and breadth of any Fort a farre off, thereby to gather, of what receite the same Fort is, or to get the width of a Riuer fronting any Fort, keeping your selfe a farre off without daunger of shot.

Chap. 22. If in a Fort or Hauen on the sea coast, or abroad on the playnes on the sea bankes, you shall see any shippes a farre off, sayling towardes you, or any Army approching by land: howe you shall alwayes be prouided in such speciall places, that in a moment almost, and with small helpe you shall know how farre they are from you, and by that meanes speedily finde when they shall be commen within the randome or point blancke of the shot.

Chap. 23. If an Army on the land, or an Nauy on the sea shall be as farre off as you may ken, making towardes your Fort. To knowe by helpe of this Familiar Staffe, how fast their gate is, and in what time they shall (according to that gate) come within reach of your shot.

Chap. 24. If standing on the sea bankes, you see your shippe of warre at the sea pursuing another shippe of the enemie, to know (by helpe of this Familiar Staffe) how farre, and how much the one

getteth

The Table.

getteth of the other in sayling, whether he be likely to ouer take the enemy, and in what time.

Chap. 25. Your selfe planted on the toppe of an high rocke, clyft, or Tower by the sea side, To knowe by this Familiar Staffe, howe deepe the leuell of the water is vnder you.

Chap. 26. If a Gunner keepe a Blockehouse, or haue a piece or two of ordinance, planted on the toppe of some very high clift by the sea side, how by this Familiar Staffe, he himselfe without any helpe at all, shal most easily and speedely in a moment, get the distance of any shippe at the sea, making towardes him, or passing a long and that most exact, for so farre as the randome of any great piece will extende.

Chap. 27. If a man were prisoner with the enemy, Howe being in the toppe of a Tower on the leades, or out of his prison windowe he might by this Familiar Staffe knowe the deepth to the grounde, to see if he were able with any deuise to let himselfe downe without daunger.

Chap. 28. If a Fort or Tower stand vpon an high hill, How by this Familiar staffe to know the ioint and seuerall heightes both of the hill and tower.

Chap. 29. If being at the sea you would cast ancor as neere some Fort or Harborrowe as you might be free from reach of their shot, How by this Familiar Staffe you shall exactly get the distance thereof, or the distance of any other shippe from your shippe beeing both fleeting at once on the wilde sea.

Chap. 30. How by this Familiar Staffe to carry the leuell of one place to any other, necessary for such as shall vndermyne a Fort, to knowe alwayes how deepe they are. Or for such as would try whether waters may be brought from one place to fortifie another.

Chap. 31. How by helpe of this Familiar Staffe, you shall cary a myne vnder the ground, and set barrelles of gunpouder direatly vnder any Tower or chiefe place of any Castle or Fort.

Chap. 32. How a Captayne may by this Familiar Staffe, set in Plat or Mappe any Prouince of the enemies Countrie.

Heere beginneth the Booke of the readie, eadie, and pleasant vse of this new Instrument, called the

FAMILIAR STAFFE.

CHAPTER. I.

What moued the Author at this time to publish so much of this Instrument and his vse, as he now setteth forth.

It was my good hap to be at the mansion place of my most honorable fauorer the right Ho. Sir Frances Knolles knight, called Greyes Court in the Countie of Oxenford in Summer last, where the right excellent and most noble Lorde, Robert Deuorax Earle of Essex, his grandson beeing expected that day, it pleased his Honor to passe the time, or rather as it might be iudged, to stirre vp by his example the couragious minded knights and Gentlemen his sonnes, naturally apt inough of themselues to patrizate, imitate, or rather to excæde in all such magnanimous exercises, there to contend with the right worshipfull and valorous Gentleman, Sir William Knolles knight, his sonne & heire, in shooting with a small pæce of ordinance at a marke, which pæce was there ready on her carriage, appointed at that time together with another great pæce, not long before the Spanish kings, called a Saker, which the sayde sir William to his high deserued commendations, then lately had brought home

B   from

The vſe of the

from the winning of ẏ Groyne in Spaine, & many other smal shot there also set readie on the leads of the house, with trains of gun pouder to be shot off, to welcome the said noble Earle: sending forth amidde the regions of the aire the exceeding ioy that was there taken of the safe and happie arriuall of his nobleneſſe, from the then late desperate voiage performed into Portingale.

At this time after his H. and his sayd sonne and heyre had each of them shot, I being at hand, was demanded a queſtion or two of getting the diſtance of the marke wherto they shot. And the sayde Sir William Knolles was also verie desirous and inquisitiue of me, in what time himselfe might learne skill sufficient for that purpose. Of which so iuſt and wished occasion I was right glad, and the rather when I considered with my selfe, how these dangerous times threaten to set our great ordinance mightily on worke, and how needfull & appertinent it is to the shooting in them certaynly and speedily, to get the diſtance of the marke they shoote at, as well for annoying of the enemie, as sparing of shootes spent in waſt. So that I was in manner assured, that no Treatise was like to be more acceptable vnto his H. (next vnto bookes of sincere religion) then such as should tend towards the defence and safe keeping of this his natiue Countrie. Wherein it is wel knowen, euen to any that leaſt know him, how forward his H. hath alwais bin and continueth euerie waie whatsoeuer. That olde age and graie hayres cannot yet daunt or holde him backe from the face of that enemie that should inuade the same, or offer violence or hurt to the person of our moſt gracious soueraign Ladie and Queene Elizabeth, whome God graunt long to raigne ouer vs to his glorie. Considering also what no small want of some one perfect meane, Generall for all these actions yet remained after so great a number of both Latine & Engliſh authors, writing of so sundrie waies to attaine Altitudes, Longitudes, Latitudes, Diſtances and Profundities, some by the Quadrate, some by the Trigõ, some by the Croſſe Staffe, some by the Cyrcle, some by the Quadrante, some by the

bare

Familiar Staffe.   3

bare rule or Squyre, some by plaine stickes and stations, some one waie some another. And amongst them all, no one instrument sufficiently apt for all purposes, neither yet to doe any of these feates in all pointes, or almost in anie point of himselfe, without the continuall & readie vse of the fiue partes of Arithmetike. Which in the vnlearned Gunner is not alwayes to be had: nor hauing it, easie for him to performe, especiallie in such busie times, when the buzzing of dangerous businesse amongst the multitude, shall bring the best learned and skilfull man out of his numbers: besides the number of preceptes so toilesome, the manifolde absurdities admitted, so loathsome. Vpon al these iust occasions I was in maner-forcibly drawen at this time to bend the fabrication and vse of this so noble instrument wholie vpon these causes, which I had for three or foure yeeres before determined to haue published in a more ample sorte, almost as generall for a number of matters of Geometrie, as my Iewel is for matters of Astronomie, Cosmographie and Nauigation, which I yet meane to doe, but that I dare make no more promises in print, vntill I haue finished the second part of my Iewell, so often since required at at my hand, being fiue yeres past: which yet this busie world suffereth me not to bring forth, although I contend to loose no manner of time.

Chap 2.

Of the imperfectiõs of the Crosse staffe called by Gemma Frisius, Baculus Geometricus, *and how much this Familiar staffe exceedeth it and all other Instrumentes hitherto deuised to like purposes, both for sufficiency and facillity.*

FOr that I haue bene acquainted with diuerse willing wits desirous of knowledge, wanting the rudiments of Geometry, whereby they might be enabled to make choise of such instruments as might be subiect to fewest errors and absurdities, to haue bene marueilously in loue with the crosse staffe

The vse of the

and that altogether, because it yeldeth his vse with that facillity, though but in matters at hand. For which cause, I thought good euen of zeale and good will to all such willinge young practisers, to waste this Cap. for their good, to assure them how weake and to small purpose the crosse staffe is, in respect to be imployed in any weighty seruice, scarce doing any thing well, and that not much aboue a bowe shute of, whereas this my familliar staffe shall performe them no lesse exacte at a mile, 2. or 3. distante, if the marke bee suche that the eie may cary it strongly.

There haue bene from time to time besids the crosse staffe diuerse other instruments, deuised for these kindes of scruisable mesurations, as in the Cap. before I mentioned: of which the Geometricall quadrate was thought to be the best, as it seemeth to me by G. Frisius, who often vpbraydeth the same quadrate with his staffe, when he had in as much as mighte be doen, reduced the crosse staffe to all perfection, letting it then to lacke no commendations, saying euen at the first entraunce to the vse of it, lib. de radio astro. & Geomet. cap. 5. Dimensionibus longitudinum altit. latit. & distantiarum antecellit radius reliqua instrumenta ad similem vsum excogitata facillitatem habens longe maiorem & copiam. But by G. Frisius good fauour, I see no cause why it should carry any of those commendations, more then for the facillity in vse, and for this, wilaske no better iudge then him selfe. For in taking of an altitude, which is simply the best thing it can doe, if the staffe be not directed leuell with the horrizon, then cannot the crosse bee parallell to the vpright, and therefore breedeth error as him selfe confesseth in the same 5. Cap. saying: Vnum a paucis notatum intollerabiles inducit errores. Si quidem Radius in dimensionibus per directam lineam, & quasi ad normam tendere debet versus lineam quam metiri statuimus, siue ea sit longitudinis siue latitudinis, what can a man aske more then direct confession, yet after he sayeth againe by way of auoidance in the same Cap. Neque vereri debes paruum

a norma

Familiar Staffe. 5

a norma deflexum qui nullum inducere potest errorem sensu perceptibilem. But by his fauour againe though a litle holding from the leuell, can breede no great error in the altitude or latitude. Yet that very smal error, there comitted: shall growe to somewhat in the longitude or distaunce: being he teacheth no other way in maner, to get a distance but by the altit. or latit. of some thing at the xtreame of the distance first obtayned, multiplyed into the length. Which in a long distance will multiply that insensible error to become intollerable as he termeth it. But what should we talke of longe distances, when as he maketh a prouiso, we shall deale with none by these wordes in his 5. Cap. or the like in effect. Nihili esse quantum distes a re mensuranda, verum opus est distantia non admodum longa. And then were wee as good to be without his staffe in my opinion, as to entertaine him on such streight conditions. But now to remedy those errors in altitudes, he sayeth in one place you may sette the staffe leuell by a thredd and plummet aplied to the crosse. In erectis quidem appenso perpendiculo. Yet in another place he thinketh that, to combersome, saying thus. Qua in re, sufficit rudius oculi indicium ac examen. As for latitudes transuerse: remedy he coulde finde none, saying: In transuersis vero, visui credere oportet. Yet for all these things he could not holde but conclude towards the ende of his first practises. Cap. 2. Itaque nulla in parte superatur a scala Geomet. verum multis modis superat, quin ipse radius, scala quoque quædam est geometrica, tanto p æstantior quanto maiorem prebet vsum, he should haue said facilorem vsum and then he had sayd somewhat, for taking away his facillity, you take away his chiefe goodnes. The quadrate, the plaine stationes which stickes, the bare rule or squire, being wel handled, will do much more, as wel in great as small distances, onely they want the like facillitie in vse. Yea and G. Frisus him selfe for all his great bostes before let fall was driuen ere he ended his booke to seeke another waye for his matters in maner quadrate wise, by hanging his crosse

staffe

Staffe, as it were by the haire of the head, with his armes abroade, and then is his so well boasted facillitye quite gone: Besides that, the longest parte of your goodly instrumente is brought then to halfe the length of the crosse, and a small length of the staffe, which in a staffe of 5. foote length will bée but 15. inches, and in a staffe of 9. or 10. foote, which no mans height can possibly reach to vse in that manner without a ladder, the longest part will be little aboue 2. foote, aud of that length though the facillity remained, a man might as easelye wilde a quadrate. And now considering that halfe the crosse and 3. quarters of the staffes length, is by this meanes abiect and to no vse, who would carry all that wood about with him, except the weather were cold. Therefore to conclude, let no man be led away with the facillity of the Crosse staffe, or of any instrument, rather let him thinke no pains too great that produceth an exacte truth. And where as G. Frisius sayeth, that his staffe is quædā scala geometrica, I say that my Staffe is not quædam, but summa scala geometrica, that he is nobilissimū trigonū geometricū. For what may be doen by Triangles in Geometry, whole volumes do testifie, and my selfe in another treatise of this staffe at some other time will partely manifest. And I say that he shalbe quoddam quadratum Geometricum and quidam radius Geometricus, and performe the vse of his Radius more effectually then it performeth the quadrates vse. What should I say, it is Baculum Catholicon siue generalissimum, and shal recompence the tractable facillity of the crosse staffe, with his familiar palpaple vse, as well in his performance of these mensurations, as his daily vsage for an ordinary and familiar walking staffe. For which cause I haue named him Baculum Familliare, the Familliar staffe.

Cap.

Chap 3.

Of the framing and fashioning of this noble instrument, called the Familiar staffe.

THis ſtaffe conſiſteth of two ſeuerall partes, or rather two ſeuerall ſtaues, the one I call the ſtanderd ſtaffe, the other the running ſtaffe. Both in manner like, very plaine to conceaue: being each but as it were, two ſtreight rulers iointed together, to open and ſhutte euen like an ordinary paire of compaſſes.

 The ſtanderd ſtaffe is here repreſented by this figure A B C, hauing his two legges or limmes A B and A C, ioyncted at A, each legge conſiſting of 3. rules, or narrow boordes 5. fot in length, of breadth and thicknes conuenient to carry it ſelfe ſtreight, of ſome ſuch wood as wil beſte keepe from warping, glewed and wrought together, ſo that both legs A B and A C being ſhutte cloſe : ther may be within it a concauity of one pnch and an halfe ſquare quite through to containe the running ſtaffe for eaſy carriage. After this, for eaſier handling in vſe, let this ſtanderd ſtaffe be brought into an 8. ſquare. And either lette the round or panell of the ioincte A be of 4. inches diametre at the leaſte, hauing a ſcrewed pinne of yron throughe the centre with a ſcrewed boxe on the vnder
                ſide

The vſe of the

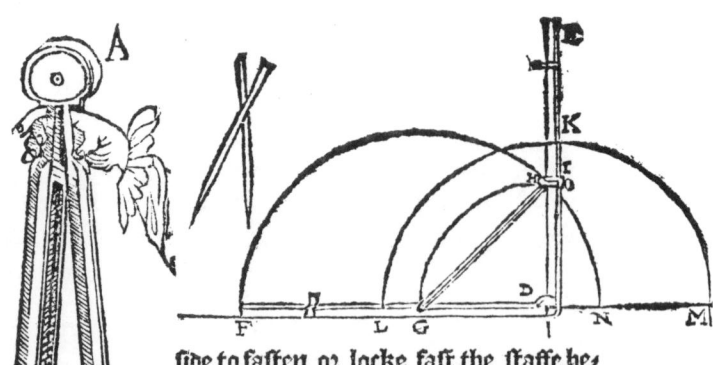

ſide to faſten oʒ locke faſt the ſtaffe being open at anie angle oʒ width : oʒ els
let ther be a third rule oʒ graduatoʒ, like
vnto that of the running ſtaffe, which is
far moʒe ſure if you may conuey him into the roome , and ſo may the ioynte
A. be as ſlender as you will. Within
the hollowneſſe of this legge are two
channelles made in the vpper and lower ſides parrallell, to their fiduciall
oʒ centre lines : by oʒ in which, two
boltes oʒ peeces of woode, vʒ. M. ʈ N. do
ſlippe to ʈ fro carrying either two ſight
pinnes halfe rounde, tʰe yʳ plaine ſides
running cloſe to the ſtaffe, euen with the
fiduciall oʒ centre lines , hauing each a
pretie button oʒ pearle on the top, thereby to direct the eie to anie marke : oʒ els
two halfe round deepe holes oʒ ſockets,
into which two like ſight pins oʒ anie
other kinde of ſightes may be put at a-
nie time when you ſhall vſe them : other
two like ſight pinnes , you muſt haue
fixed neere vnto the centre, their flatte
halfe rounde ſides euen alſo with the fiduciall lines , on each
leg

Familiar Staffe.   9

leg one. The vse two bolts or péeces carrying the two running sight pins must haue each of them a spring of mettall, to cause them runne the more pleasant, & also a screwpin at all times to fasten them at anie part of the standardeslegge. So much for the standard staffe.

The running staffe represented in this figure by E.D.F. is ẏ verie like to the standard, being also but the like two legs or limmes D.E. and D.F. ioynted at D. each made of thrée rules or narrowe boordes of foure foote length and better, and of breadth and thicknesse, conuenient to fill the hollownes of the standard staffe, hollow as the former, not eight square, but foure square. Herein it differeth, that at the point G. of the side D.E. twentie inches from D. the parte G.F. shall bee to take off and on with helpe of a square socket or some screwe, made in the end G. of the parte T.G. to be taken off, when it is to be vsed as ẏ Guners quadrant, or to be set on at E. to inlarge the side D.E. as occasion shal serue. Also at the point G. there must be ioynted another light foure square hollowe ruler, viz. G.H. equall in length to G.F. which (the instrument shut together) must fall into the channels of G.F. and E.I. This square ruler G.H. I call the Graduator, because hee yeeldeth degrees. The centre of this graduators ioynt G. shal be euen with the middest of the breadth, and his length shal be such, that taking D.I. in the side D.E. iust equall to D.G. and the running Staffe opened exactly to a square angle, as here you see, that the extreame of the graduator, viz. H. which I will hence forth cal his Apex, may but iustly touch the point I. making there a right angled Æquicrurall triangle. And let the graduator haue at his Apex H. some deuise with a socket or payre of chéekes, or else annexed to some such running bolt, as before in the standard carrieth the sightes, and a screw pin to fasten him in anie parte of the side.

There must be in the centre D. a pinne turning round, hauing either an halfe round hole, quite thorough, to fit vpon the sight pins which the running bolts of the standard do carrie, or else an halfe round pin standing out to fit into the halfe

C   round

The vſe of the

round holes oꝛ ſockets of thoſe boltes. You muſt haue alſo foꝛ each ſide of this running Staffe, one running ſight oꝛ wing (as they call it) like vnto thoſe of the Croſſe Staffe, as in this figure you may ſee, & alſo two fixed ſight pins foꝛ eache end of

A Supporter.

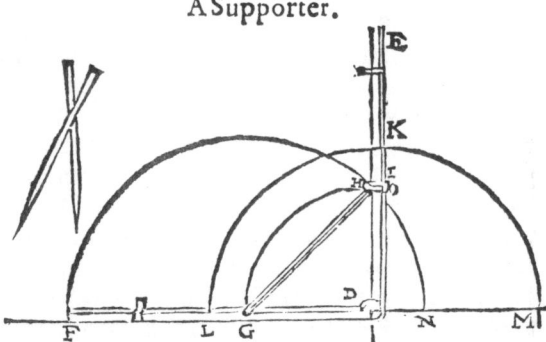

each ſide one parallel to the fiducial lines, which foꝛ diſtinction ſake, I will call ſight pegs: and muſt be ſo planted, that they may riſe and fall into the ſides, ſuffering the wings to ſlip ouer them when occaſion ſerueth. And if the ſtandard it ſelfe had on each legge the like paire of ſight pegs, it were not amiſſe.

Your inſtrument thus pꝛepared neatly in all pointes, as a woꝛkman can better of himſelfe tell how, then if I ſhoulde wꝛite much moꝛe. You ſhal diuide each legge of either of theſe ſtaues into as many inches as they will containe, beginning from the centres A. and D. and euerie inch, into eight partes, which partes ſhall ſignifie feete, paces, perches, pike lengths, oꝛ anie other kinde of meaſure that you ſhall vſe: ſo ſhall the fiue foote length of the legs A.B. and B.C. of the ſtandarde, containe each 480. equall partes, and the ſides D.E. and D.F. of the running Staffe as many like partes as it may containe, all which to be ſubdiuided againe, each into foure equall partes, and ſet numbers vnto them, encreaſing by tens, each ſoꝛte of diuiſion dꝛawen to his ſeuerall ſpace limbe-like, as vſually woꝛkemen can doe full well. Thus is your inſtrument fully

Familiar Staffe.   11

fully finiſhed, ſauing for his degrees, which in the next Chapter I will ſhew.

Other neceſſaries there are appertinent, as a threed and plommet for the running Staffe, or els in ſtead thereof (which is much better both for vſe and carriage) ſome pretie rule of mettall, equall in length to D.G. heauy towardes the lower end, mouing pliant, ſometimes on the centre D, and ſomtimes on a pinne ſet-on the hanging ſide of the ſtaffe, thereby at anie time to ſet him perpendicular to the Horizon. A copartner at all times to helpe, for that the inſtrument is great, a couple of ſupporters, if you will, which neede not be curious, they may be made of two ſtiffe ſtickes three or foure foote long, nayled together about three or foure inches from the vpper ende, ſo that they may open and ſhut vpon the nayle, making a forke aboue to reſt the ſtaffe in, as in this figure you ſee, or any other deuiſe you like better. But if anie man deſire to haue this Familiar Staffe made in a leſſe or more ſlender proportion then this before deſcribed, thereby to be more neate and fit for a dayly walking ſtaffe. Then let him haue no running ſtaffe at all, but ſet the graduator in the ſtandard it ſelfe, and haue a ſmall thyrd foure ſquare ruler, with a centre hole, in all reſpects like the one ſide of the running Staffe, looſe, to vſe at pleaſure. For ſo ſhall it be ſufficient enough for anie purpoſe, onely the running Staffe helpeth to doe all things with more ſpeede and facilitie, which two ſtandards alſo made in this ſorte may as well performe. For the one may ſerue in ſteade of the ſtandard, the other as the running Staffe.

Cap. 4.

How you ſhall in a ſingular ſorte ſet degrees, on the leuell ſide of the running ſtaffe, and alſo on his Graduato, together with the points of the gunners quadrant.

Hauing had the inuention and vſe of a geometricall inſtrument not much vnlike this my Familiar ſtaffe, nere theſe

C 2   7. yeares

The vse of the

7. yeares, but of much smaller size then my good wil desired, because I could neuer conclude with my selfe, how to compas conueniently without comber, the degrées of the circle thereon: neither yet what foot or rest to cary him on, in vsing him, vntill now this suddeine occasion ministred by my honorable frends in the 1. Chap. mentioned, did new sett a sharpe edge on my dull witts: yea although I conferred sondry times with diuerse mathematically geuen about the same. As for the foote or reste which I coulde neuer deuise but very great and more combersome then the staffe it selfe, requiring a spare body to cary it: Now I meane that spare body shall supplye the vse or stead thereof: as heareafter shall appeare, but for the graduacion you shall thus after a singular sorte supply.

The running Staffe framed, as in the last Chap. I wil call the side D E (whereon the graduators apex must run to and fro) the leuel side, because in al questions for the most part it must be carried leuell with the horizon. The side D F I will call the hanging side, because in euery leuell the plum line must hang directly by it: the piece G F being taken from the side D F I will call the iointe piece, this knowen: you shall on some plain boorde of two foote bredth draw a ground line, vz. M F then on some point thereof, vz. D erecte the perpend: D K of 2. foote length at the least: Wherein you shall sette the pointe I. xx. inches iuste from D, and likewise G as much from D, in the groundline D F, then on D with the quantity D K, describe the semy circle M K L and on G with the quantity O I, describe the semy circle N I F. This doen deuide the semy circle M K L into 180. degrees, and euerye degree into sixe partes at the least, with blinde notes: you shall also on the centre D with the quantity D I describe the quadrante I G, which you shall deuide into 12. equall partes, which are called the 12. pointes of the gunners quadrante, and euery of those 12. partes, againe you shall subdiuide into other 12. partes, which are called the minutes of a gunners quadrante, so haue you 144. partes.

Thus is your ground plot laide, whereby to set on all your graduations.

graduations. Wherefore you shall now bring your runninge Staffe prepared as in the last Chap. & opening it to a square angle, (which is doen when the apex of the graduator is fixed to the pointe I of the leuell side) place him now on this figure his centre D, on the point D of this figure, and so consequently I vpon I, G vpon G, F vpon F &c.

The tipe of graduating this instrument, commended by the Author, to the ingenious matthematicall minded gentleman, Master Auditor Hill.

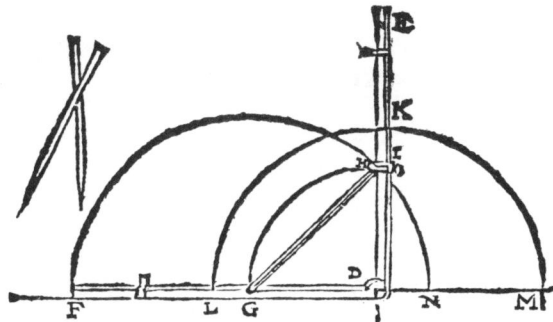

The vppermost side of the graduator you shall deuide into 90 degrees, by a rule laide from D on euery diuision of the quadrante K L, and the vndermoste side into the 1 4 4 partes of the gunners quadrante, by the quadrante I G in like manner: So is your graduator furnished with his partes, yf you draw certaine parallell lines, limbe like as the manner is, and adde but numbers to each of them: but now to set graduation in the leuell side D E you shall keepe the hanging side D F, sure fixed on the groundlyne M F keeping his former place, & bring the leuell side euen with it, shutting in the graduator close out of sight: then shall you open him againe till his fiduciall line doe cut one degree aboue L of the semicircle M K L, and there lifte vp the graduator til his apex doe touch the leuel side.

ſide. Marke that touch point, foꝛ one degrée: then open the leuell ſide vnto 2. degrées aboue L, and there bꝛing the graduatoꝛs aper againe to touche, marke that touche pointe in the leuel ſide, foꝛ 2. degrées, and ſo doe from degrée to degrée of the whole ſemicircle M K L, marking ſtill the touch points in the leuell ſide, and you ſhall finde that the aper will touch that leuell ſide euer in the ſemicircle N I F but when you are once paſte 10. oꝛ 15. degrées, you ſhall make notes in like manner foꝛ euery ſubdiuiſion of eache degrée, and let theſe degrees be placed either in the vpper oꝛ lower flatte of the leuell ſide, it is no great matter which, and numbꝛed from E ending at D with 180. degres, foꝛ note, that the euen partes mentioned in the laſt Chap. muſt be ſet on the inner oꝛ ſhutting flats of both ſides of this running Staffe.

Chap 5.

How by this Familiar Staffe to trie whether a peece of ground be leuell, whereon to plant your peece of ordinance.

Of leuelling. I Perceiued by that ſmall pꝛactiſe I ſawe in ſhoting, mentioned in the firſt Chapter, that it is needfull to place a peece of Oꝛdinance to be ſhot, on a leuell plot of ground, ẏ one wheele of theyꝛ carriage goe not higher then another, leaſt in her recoile it make her ſhoote awꝛie: firſt ſticke ſome wooden pin in the ground, and reſting the centre of your running ſtaffe, (ſet at the ſquare angle) theron: lifting vp the hanging ſide, by helpe of a thꝛeed and plommet oꝛ plum rule, vntill the leuell ſide be exactly placed leuell with the Hoꝛizon: then ſhall you from that wooden pin extend a thꝛeed oꝛ line, euen by that leuell ſide of the Staffe ſo placed, and with another pinne alſo thꝛuſt into the ground, there faſten it, then keeping ſtill the centre of your ſtaffe on his firſt pinne, turne about the leuel ſide ſomewhat wide from the line you haue ſtrained between the two pins, and there in the very like manner place another thꝛeed oꝛ line extended from the ſaid firſt pin by the leuell ſide

FamilliStaffe. ar 15

of the Staffe, and let the same be holden stiffe out with a third pinne: now laying your eie nere the ground, by these two lines or threeds so extended, you shal see whether any part thereof be higher then the leuell, to bee pared awaie, or lower, to bee filled up E. A. N. is the running Staffe A. the first pin K. the second, I. the third A. K. and A. I. the two threds or lines extended leuell with the Horizon P. the eie leuelling by the threds, O. an hillocke to be pared awaie.

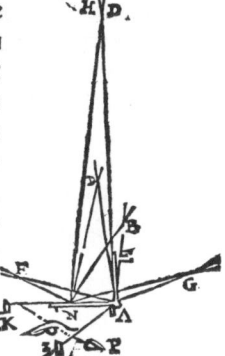

Chap. 6.

How by this Familliar Staffe to mount a peece of ordinance by pointes of the Gunners quadrate.

HAuing by the last chap. leuelled the ground, whereon your *Of mounting* ordinance is to be planted, next followeth to know how to mount your peece unto any point desired, which is most easie: For your running Staffe set at this square, pulling awaie the ioynt peece of the hanging side, so is that side the shorter by more then halfe. Then putting his leuell side into the peece up to the verie aper of the graduator, there let it bee holden euen and leuell, with the bottome or lowest side of the conca-

A peece leuelled at point blanke.

uitie of the peece, then lifting up or weigh-
ing downe the peece till the threed and
plummet, or plumkin rule, hanged out of
the centre doe rest euen with the fiduciall
line of the hanging side, ther is your peece scituated leuel with the Horizon neither mounted nor embased anie point or minute, but euen at point blanke, as they call it. After this, if the peece be eleuated neuer so little aboue that leuell, then doeth
your

16   **The vſe of the**

your thred and plummet or plum rule, ſhew on your gunners ſcale (appointed in the fourth Chapter, on the inner ſide of the graduator) to what point or minute the peece is eleuated for to ſhoote at random, as they call it, yea, and if need be, it ſheweth you on the vpper ſcale of the graduator, the ſame random

A peece mounted to ſhoote at random.

by degrées of the circle, which for anie thing I can ſee, might be applied to this purpoſe as wel as the gunners points.

Chap 7.
By vvhat meanes a certaine table is to bee made, thereby to knowe hovv farr any peece vvill ſhoote at random, beinge mounted to any poinct of the gunners quadrante.

Let no man thinke that I pretend to ſhewe the arte of ſhooting in great ordinance, being in deede a matter that I neuer tooke occaſion ſo much as to thinke of, before this time, but for that matter, do aduiſe euery man to repair vnto Tartaglia, which is lately engliſhed by M. Cyprian Lucar, & publiſhed. Who hath ſet down diuerſe ſingular conceites & obſeruances about the ſame, only to the end to ſharpen your taſt for the fruit of this my worke following. I thought good to preſent you with this inſtance out of Tartaglia, which in the next Chapter I meane alſo to applie to my Staffe. He giueth but a glance in his firſt booke, 1. Colloquie, that anie one peece being mounted from point to point, and minute to minute of the gunners quadrate, and the randomes or diſtances in paces of that peece taken at euerie ſeuerall mounting, and noted in a table (being carefully dealt in) would ſerue to know the randoms of any other péece likewiſe eleuated from point to point: if you do but know anie one ſhoot at anie one point eleuated of that other peece. For looke what proportion the ſhoot of any point of the firſt, beareth vnto any other point of the ſame peece. The like ſhall the randomes of anie two like pointes of the ſecond beare one to another.

                                                      As

Familiar Staffe.   17

As for example: If the first péece shoote fortie paces at the first point, fiftie at the second, sixtie at the third, &c. Then if you shoote one shoote in another péece, which at the first point happily shooteth but thirtie fiue paces, and that by this one shoote knowne of this second péece, you would before hand know how farre he should shoote being eleuated to the third point. In this case you must by the rule of thrée (saith Tartaglia) séeke out a number in that proportion to thirtie fiue, as sixtie is to fortie, which you shall finde 52. ¼. paces: and so much shall this second péece shoote, eleuated to the third point, if that Instaunce of Tartagliaes be true in it selfe. Although many secret accidents besides may much alter it, as that the powder at each shoot be like good & like much, the pellet of like waight, the péece of one temper, and such like. For Tartaglia writeth amongst many other experiments singular well by him discussed, which I am not any way to meddle with, that a péece being cold, shooteth not so farre at the first shoote, as with like charge hée doth at the second or third being then temperatly hote: And after when she is ouer hote with shooting, she beginneth to shoote shorter againe: and sheweth the reasons thereof.

Chap. 8.

Hauing made a table of randoms to some one peece according to the precepts in the last Chapter. How by this familiar staffe, to make the same table serue for any other peece without any arithmetycall calculation.

Truely this inuention is very new: for it came into my head but euen as I was writing the last chapter, where you are taught Arithmetically to performe the same. But my familiar staffe will not haue you troubled with numbers in such daungerous times. Wherefore hauing your table made by some one péece exquisitly shot in, and one shoote giuen of some other péece at some one certaine point or minute perfectly made. So haue you alwaies, as in the last

Of proportion.

D   Chapter

18   **The vſe of the**

Chapter is ſhewed three numbers giuen: two out of your table: and the third is the ſhoote of the péece in action. You ſhall ſeeke the one of your two table numbers (it is no matter which) in the right legge of the ſtanderd ſtaffe, and thereto by helpe of the mouing ſight pin, place the centre of the running ſtaffe, ſet at his ſquare angle, or any other leſſe angle it is not materiall what: and direct the leuell ſide vnto the centre of the ſtanderd: There hold him till ye haue pulled too or thruſt out the left legge of the ſtanderd, ſo that it cut in the hanging ſide of the running ſtaffe your other table number. There wreſt the ſtanderd faſt at that angle by his centre ſkrew. Laſtly, ſeeke in the ſaid right legge of the ſtanderd the number of paces of the ſhoote giuen, and thither by helpe of the ſaid ſight pin, remoue the centre of your running ſtaffe, keeping his former angle, and his ſaid leuell ſide, directly ſtill with the centre of the ſtanderd, and queſtionleſſe the left legge of the ſtanderd ſhall there cut off in the hanging ſide of the running ſtaffe, the iuſt number of paces deſired.

    For example, it is admitted in the laſt Chapter, that the ſhoote of a peece eleuated vnto the firſt point was thirtie fiue paces, by which I would know how farre that peece would ſhoote, being mounted at the third point: I ſeeke in the Table there mentioned for ẏ ſame firſt point, there according to ẏ laſt chapter I finde fortye paces, and alſo for the third point I finde ſirtie paces, which had, I ſeeke the biggeſt of thoſe table numbers, viz. 60. on the right legge of the ſtanderd: it endeth at B there I place the centre of the running

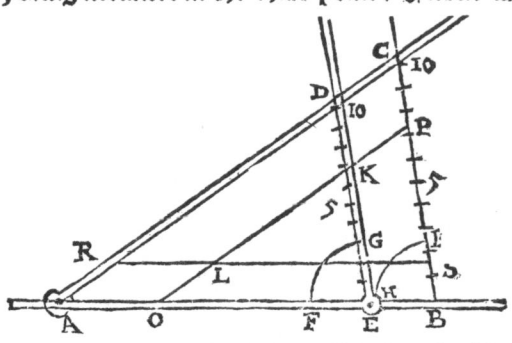

Familiar Staffe. 19

running staffe, being set at a venture at the angle C B A and apply his leuell side B A euen with the centre A of the standerd, and there on his hanging side B C I seeke the other table number, that is fortie, at C. thither I bring the left legge of the standerd to cut, and there locke him fast to kéepe the angle C A B. That done I seeke the number of paces of the shoote giuen, that is to say thirtie fiue, on the said right legge of the standerd, that is to say at E and thither I remoue the centre of the running staffe his leuell side, kéeping still the centre A and in both the staues their angles before set C A B & A B C vnstirred, and there doe I sée the left legge of the standerd to cut off in the running staffe at D fiftie two ½ paces, the Random of this second péece at the third point desired: in all respectes the same which in the last Chapter you found by Arithmetike. So that vpon the matter you haue now learned the rule of thrée by Geometry, and is demonstrated in Eucl. lib. 6. Propo. 10.

Chap. 9.

If a Wall or Tower were to be scaled, and that you may come vnto the base of it without daunger, How by this familiar staffe spedily to get the height thereof, thereby to make your scaling ladders according.

IT hapneth often that it is thought more conuenient to win a towne or fort by scaling with ladders and such like, then to batter it with ordinaunce. In such a case where the walles may be come to, without daunger, it is so easie and common to get the height by any instrument, that I could scarse spare it a roome, but that it may serue as a ground, or first introduction to the rest. This my familliar staffe as in the 2. chap. I saide, may be vsed in most cases, eyther as himselfe, or as the quadrante or circle, or as the quadrate, or as the crosse staffe, or as the gunners quadrant: and in this case indifferent to them all. For if you take your running staffe set at his square angle

Of altitude

D 2

20 # The vſe of the

angle, and apply the leuell ſide to your eye, lifting vp the centre till the line and plummet out of the centre do hange iuſtly on the 45. deg. or 6. gunners point of the graduator, which are both in one ſtreight line, what vpright ſoeuer you ſée, then euen with the leuell ſide the height thereof is equall to the diſtaunce from you. In ſo much that when at any time you ſhall by this inſtrument or any other finde the altitude of the ſunne 45. deg. if then you runne and meaſure the ſhaddow of that tower or wall, the ſame ſhall be equall to the height thereof, as moſt auchors affirme, but in very déede it ſhall be leſſe then the height by ſo much as one quarter of a degrée cometh to, as in the treatiſe of my new Aſtronomicall ſtaffe not yet extant ſhall appeare: in the meane time take the ſhaddow at 44 ¾ degrées high, and the ſame ſhall be your deſire.

But now, if you will doe it croſſeſtaffe wiſe, then ſet the extreme of the leuell ſide of the running ſtaffe to your eye carrying it ſquare to the vpright, as néere as you can, the hanging ſide vpwards and parallell to the wall, and go in or out ſo, till you may ſée the toppe of the vpright, euen with the toppe of the hanging ſide, and there ſhall you finde the diſtaunce to the

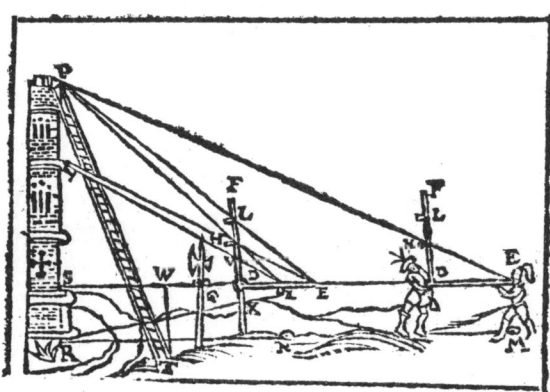

wall equall to the height thereof, as before. And if you place the wing in the midweſt of the hanging ſide, and by that ſée the
toppe

Familiar Staffe.   21

toppe from the extreme of the leuel side, then is the distance to the vpright double to the height, if at a fourth part from the centre, then is the distance 4. times the height, &c. For example in this figure the running staffe set at his square angle, and carried square to the wall as you see: if you go in or out till you might see the toppe P of the tower R P euen by the extreames E and F there shall the distance from your standing to the tower be equall to P S, then adding thereto your height viz. R S, you haue the height desired. And if you chuse your station so that you see P by the extreame E of the leuel side and the wing H: and that H be one 4 part of D F from D, then is E S foure times the height P S, because D H is foure times in D E: in like manner H being set at halfe or a third part, &c.

Chap. 10.

To performe the last cap. Where you dare not come neere the base of the tower for daunger of shot or let by reason of some deepe mote or ditch.

I Will first in this Chap. shewe you how by this familiar staffe to do this feate after the very manner almost of the Crosse-staffe, which many men affect so well, and in the next chap. after his owne fashion, and both may as well be applied to a wall or vpright approchable. *Of altitudes*

Take your running staffe set at his square angle: and place the wing of his hanging side neere about the middest thereof, or as occasion shall serue: fixed: then applying your eye to the extreame of the leuell side, directed right against the wall or tower in manner of the last chap. chusing you a station somewhat farre from the wal, where from thextreame of the leuell side you may see the toppe of the wall, by the sayd wing before fixed: which done, then shall you place the running sight of the leuell side from his extreame towards the Centre, halfe or a quarter, or iust so much as the sayde wing before fixed in the hanging side is placed from the Centre of the staffe, and there

D 3   let

The vſe of the

let this wing alſo be fixed. Then chuſe you a ſecond ſtation (going in towards the wall) carrying the inſtrument plum as before, till you may againe ſee the toppe thereof by both theſe fixed wings. Laſtly, meaſure the diſtance betweene both your ſtations for the ſame ſhall be eyther the whole, the halfe, or quarter of the height of the wall you ſeeke, according as you did ſet the laſt wing from thextreame of the leuell ſide, eyther the whole, the halfe, or quarter the length of the firſt wing from the Centre, adding thereto when you haue doone the height of your eye aboue the baſe of the vpright.

For example, Admit you placed the wing of the hanging ſide 120. euen partes from the centre of the ſtaffe, viz. at H by which at the firſt ſtation viz. M and thextreame of the leuell ſide of your ſtaffe, viz. E you ſée the toppe of the wall of the tower, viz. P by the viſiall line E P. That done, then place the wing of the leuell ſide E D one halfe of D H (being 120. parts as is ſayd) viz. 60. euen partes from E at I, and ſéeke you a ſtanding nearer to the tower P R, where you may euen by the winges H and I ſée P the toppe thereof againe viz. at N,

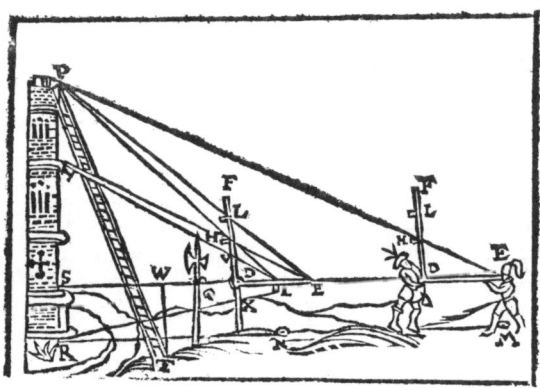

then meaſuring the diſtance betwéene your ſtations M and N which admit to be 20. fote. I conclude becauſe D H 120. parts is twiſe ſo much as E I 60. parts that the height of the tower
from

Familliar Staffe. 23

from the toppe vnto the height or leuell of mine eye, viz. P S is twise 20. foote, viz. fortie foote: and if néede be you may vse the helpe of the 5. or 6. note of the 15. chap. following.

But here I can no longer delay, but warne you of two things in this working, both which, G. Frisius halfe winckketh at in the vse of his Crosse Staffe, because he would haue it commended forsooth for facillity. But I haue euer detested that facillity that should produce any absurdity. The one is, that the hanging side of our instrument be parallel to the wal, the other, that your eye in both stations be sett in one leuell, that is to say, no higher aboue the base in the one station, then in the other, both which are fowle faults to be admitted, if the vpright should be far of: yet are easely salued with a little paines. And I earnestly exhort euery practiser in these actions, to defy that ease, that should shame him selfe & bringe his arte in question: a notable faulte amongst moste of our land measurers in these daies, I meane those that measure by platt, who partly by want of skill in Geometry, partly by want of perseuerance and knowledge in other sciences and faculties, wherwith a perfect surueior ought to bee adorned and furnished: and chiefly by want of industry, paines and diligence, doe bring them selues and a most excellent Science into disdaine and obloquy, when as seldome two of them can produce one measure or like plat of one self groūd, and that which more is, scarce one of them shal agre with him selfe. I meane produce at his second measurement either the same platt or the same measure which he did at the first. But God willing when so euer I shall goe in hand to write againe of this instrument, I shall sett downe moste certaine salues for euery sore in those cases. In the meane time, to retourne vnto the two errors, that might accreaue to this matter we haue in hand: they are easely salued with no pains at all in respect: because the greatnes of the Staffe requireth an assistaunte or partner at the further end: who shal discharge you thereof. The first is easie remedied by a plum line, or rather a plum rule mentioned in the third Chapter, applied to

the

the hanging side, as in the second Chapter was saide, which your partner may easely guide whiles he stayeth theend for you, especially hauing an halberd pitched in the ground to help him rest his hand steddy. The second as easy salued by a spare halberd, or a staffe pitched somewhat farther in towards your vpright then your second station shall be: which halberd shall haue a bright marke viz. Q placed on it, euen with your eie at the first station, by which marke Q you and your partner may easely direct the leuel side of your instrument at both stations of one height. Note that whatsoeuer is before written of altitudes, the same doth G. Frisius, Chapter seuenth de Rad. Geo. conuert to the latitudes or distaunces betwéen two markes, towers or trées, willing you to beare the crosse aside, and not vpright, and to take care that you stand square to the latitude as néere as you can, saying, a litle awry will doe but a litle harme. You may no doubt do it as well by this Familiar Staffe, as by the Crosse Staffe: but I will not bid you to do so much as a little harme, leste you doe a litle more, and then all starke naught. I meane to teach you a better way in my 21. Chapter.

Chap. 11.

How by this Familiar Staffe to performe the last Chapter another way, more exacte for long distaunces, the more safelye to keepe you out of daungerof the shotte of the forte, vvhiles you are in action.

Of altitude. Hitherto we haue but as it were borrowed the precepts of the Crosse Staffe to sitt the fancy of some, who I hope will not deny but that this my Familliar Staffe performeth them with no lesse facillity, and no great alteration: yet as in the last Chapter I sayd, my earnest care of exacte trothes will not suffer me to aduise you to relye vpon them, for anye altitude aboue a bowe shoote of, as in the second Chapter I touched, for many causes too long here to recite. But in

case

Familiar Staffe.   25

caſe the vpright be far from you, whoſe height you ſeeke, then ſhall the proper and naturall working of this my Familiar Staffe ſerue your turne much better in this manner. You ſhall firſt by helpe of two ſtationes as in the fifteenth, ſeuenteenth or twenty Chapter following is taught, get the diſtaunce from you to the baſe of the wall or tower, viz. howe many féete, yards, paces, or other meaſure, it is from your firſt ſtation: thoſe reckon on the leuell ſide of your runninge Staffe from the centre, and thereto place the wing fixed: that one, ſet that wing to your eie (placed iuſt ouer your ſtation) directing the centre of your Staffe to the wall, and let your partener kéepe the hanging ſide plumme by the plum rule, and with al, be reddy with his hand to moue the wing of the hanging ſide, by direction of your eie, ſo much aboue the centre till your ſelfe may from the ſayd winge at your eie, by this ſecond wing ſée the toppe of the wall or tower. Then ſée what euen parts are contained betwéene the centre and this ſecond wing on the hanging ſide, for ſo many féete high is the toppe of the tower aboue the leuell of your eie.

For example, admitte you had by the fifteenth, ſeuenteenth or twenty Chapter following, found the diſtance from the ſtation N in this figure, vnto the pointe of the towre leuel with
                                   E   your

76 The vſe of the

your eie viz. S. to be an hundreth pace or fiue hundreth foote, all is one, but that in getting the altitude of any thing, the reckoning had néede to be kept in féete, though for distances or longitudes, paces, perches, pike-lengthes, or any other may ſuffice: thoſe hundred paces I reckon on the leuell ſide of the Staffe, they end at I. where I place the wing.   Then directing the centre of my Staffe, viz. D. to the towre or wal, I cauſe my partner to lifte vp the wing of the hanging ſide, vntill it come directly betwéene I and the towre top, viz. at H then I looke on the Staffe what euen partes I finde betwéen the ſighte H and the centre D, admit I finde eight.   Therefore I conclude, that wall or tower to be eight paces highe, which is forty foote, and in this working you ſhall take helpe if néede bée of the fifte or ſixe notes of the fiftéenth Chapter following.

Chap. 12.
How you ſhall know by this familiar ſtaffe, the depth vnder the iuſt leuell of your eye, of the baſe of any tower vnaprochable when the ſame baſe is to be ſeene.

Of profun-
ditte.

SVch may be the aduantage of the ground whereon wee make our ſtations, that the leuell of our eye may be much aboue the baſe, and then are we not ſatiſfied for the height of the vpright, if it be vnaprochable: wherefore when you haue by the laſt Chapter gotten the height of the wall or towre aboue the leuell of your eye: note then ſome marke on the wall, euen with that leuell: and ſo may you as eaſily and by the ſelfe ſame meanes get the depth of the baſe vnder the leuell, as in the laſt Chapter you got the height of the toppe aboue the leuell.   There is no difference in the worke, but that you muſt turne the extreame F of the hanging ſide of the ſtaffe downewards, and ſet the wing thereof iuſtly betwéene the wing I, and R the baſe of the towre, that is to ſay at X as in the laſt Chapter you did ſet it at H betweene I and the top of the towre P, and the partes of the ſtaffe betweene the centre

Familliar Staffe. 27

tre D and the wing at X shall be your desire. Which added to the former height P S maketh vp the full height from the toppe to the base, that is P R.

But where the ground hath no great aduantage of the vpright, I would neuer wish you to make two workes of one, for if it be leuell with the base of the towre, it is a small matter to lay your selfe downe and worke the last Chapter close by the ground: if it be lower then the base, then place the leuell of your eye no more aboue it, then that it may direct to the base: such like small matters he that could not of himselfe prouide for, I would not haue him allowed for an executioner of these seruices.

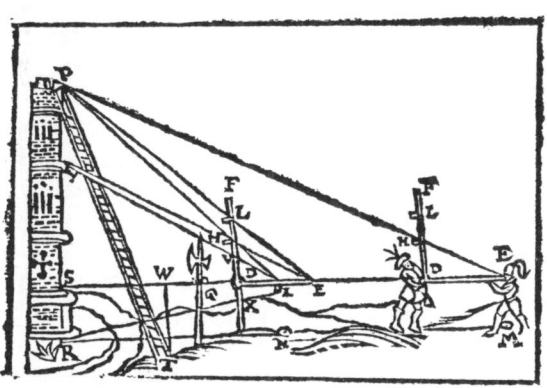

Note, that if you list not in this working to turne the hanging side of your staffe downewards, or that the ground will not admit it to sinke so low as the leuell of your eye marked on the towre, you may worke it as well and easily, kéeping the hanging side vpwards, in such maner as in the twentie sixe Chapter following is taught: and then must you kéepe the centre of the staffe ouer the station N and direct the extreme E of the leuell side towardes the towre or vpright, as by the bare inspection of the triangle I V O of the figure of the same twentie sixe Chapter: you may easilye con-

E 2 ceaue

ceaue. The quantitie I.V. of the staffe, shewing the depth of the point C vnder X the leuell of the eye in that figure as here the quantitie D X sheweth the depth of the point R vnder the point S leuell with the eye in this figure.

Chap. 13.
To knowe the length of the scaling ladder to reach ouer the ditch to the toppe of the wall or tower.

The height of the wall or tower together with the distance thereof from your station first had by the 11. and 15. chapters and helpe of the 12. then measure from that station directly towardes the tower till you come at the out side of the ditch where the fote of the ladder must stande, take that measure out of the whole distance, keepe the remaine, and then as in the 30. chap. shall be shewed, proue with your leuell whether the footing of the ladder be no higher or lower then the base of the wall, if it be higher, then take it out of the height, if lower, adde it thereto, and keepe that number also. Nowe shall you reckon the first number kept on the leuell side of your staffe, set at his square angle, and the second number kept on the hanging side, and there make notes then applying the ioyntpeise or any part of the standerd betweene those notes you may thereon tell with a sticke as they say, howe many foote long your ladder shall be.

For example in the figure of the 11. chap. T is the footing of the ladder, M the station, M T measured and taken out of M R leaueth R T the first number kept, this point T is so far vnder the leuell of the eye, as W T or S R commeth to, therefore adde it to P S, so haue you R S your second number kept, and those two numbers reckoned on the seuerall sides of the running staffe as is sayd, and the ioyntpeise layde betweene them yeildeth you the Hypothenusa P T which must needes be the length of the scaling ladder desired.

Chap. 14.
How at any station either by the standerd or running staffe the angle

Familliar Staffe.   29

angle of station or position betweene any two markes or places is to be taken two seuerall wayes.

Of taking angles.

I Am not ignorant that an angle of position in Geography is the angle comprehended betweene the visuall line directed vnto any place appointed, and the meridian line of your being. Bet if we call the angle made by two lines issuing from one place or station vnto other two places, an angle of position (though happily more aptly to be termed an angle of station) the matter is not great, being it is the angle of position or placing of the one place from the other in respect of your station. This angle shall most easely by this familiar staffe two waies be taken, the first easie and familiar to conceaue, the second more easie and familiar to be performed.

For the first way suppose P your place of station and N and T the two places or markes whose angle of station or position you seeke. Let your partner and you get ech of you your supporter mentioned in the 3. chap. on which two supporters, you shall so rest the one legge of your staffe, viz. A C. (his centre iust ouer your station marke, viz. ouer P ꝑ you may by the sight pinnes, or pegges of that legge see the one of your places, viz. N:

E 3   there

there looke that leg of your staffe lye fast on his rest, leauing that to your partners charge by his steddy holding of his supporter both with hand and knee. That legge so setled euen with the mark N, you shall gentlye open the other legge A B lifting it vp and downe withall, vntill by the two toppes of the sight pinnes of that legge A B you may see the other mark T, and there with your centre screw pinne, locke your staffe fast at that angle T A N: for the same is the angle of station or position desired. And if you take this angle by the running staffe in this manner, then doth the graduators apex also shew you the degrees which that angle contayneth on the leuell side of the staffe. But for your supporters, it is not materiall to make such reckoning of them as to be choise about them: for two stickes pulled out of an hedge, knit together with a point towardes the one ende may serue for a shift. Or if a Captayne haue but his halberd, let him sticke it fast in the ground vpright, and with his garter tye the one ende of the legge of the staffe thereto, he may then vse the other at his pleasure without helpe. Or sometimes he may rest him to a tree, or on the very barrell of the Cannon. In these cases I would haue no man tyed to precepts, but be ready to supply as occasion shall minister.

Now for the second wale, let your station be P. and let L. & T. be the two places or markes, whose angle of position you seeke. Sticke your dagger or anie stick vpright in the station marke P. in the ground, and thereon rest the centre, viz. A. of the standard or running staffe. Then applie your eie vnto the extreame of his one legge, viz. to B. and let your partner apply his eie to the extreame of the other leg, viz. to C. each of you opening his leg by little and little so wide, till your selfe see the mark T.

Familiar Staffe.   31

by the sight pins
or pegs of the leg
A.B. & your part-
ner sée the marke
L. by the sights of
the legge A.C. &
there wreast fast
your Staffe at
that angle C A.B.
for the same is e-
quall to the angle
T.A.L. by Eucly.
lib. 1. propos. 15.
which is the an-
gle of position be-
twéene the marks
T. and L. desired.
And if you perfor-
med it by the run-
ning staffe, then
shall the gradua-

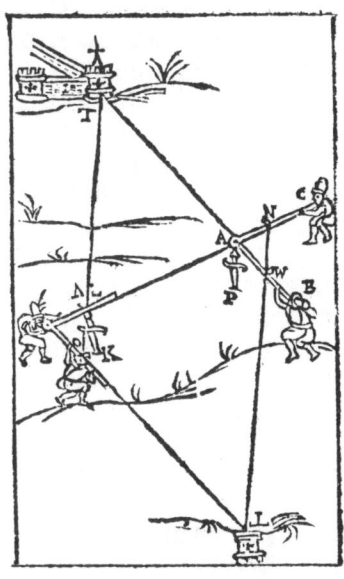

tors aper, as before among the degrees or graduation of the leuell side, shew you also what that angle is, and
how many degrées of the circle it containeth. Yea, your partner and you may as well and more speedily doe it without a-
nie dagger or rest, but that a rest may happilie guide the centre of the staffe iust ouer the marke. Note also, that in taking
the angle in this manner by the running staffe, let the leuell
side be first directed to his marke, and let him that directeth it,
pull the aper of the graduator to and fro, till his partner finde
the hanging side opened euen with the other marke.

Chap. 15.

*How (in manner of the first and plainest meanes, mentioned in
the last Chapter to take the angle of position) to get the most
exact*

The vſe of the

exact diſtance of a Caſtle or Fort from you, yea though the ſame Fort be tvvo or three miles off or more, vvhereby you may knovv hovv to place your maine battel as neere as may be vvithout danger of ſhot from the Fort, and alſo in vvhat ſpace you may march to the ſame vvhen you vvill.

Of longitude or diſtance.

ALL this while we haue about altitudes reckoned vppon feete or paces at the greateſt, but now ſhall we for longitudes or diſtances, deale (& that as certainly) with paces, perches, pike lengths, or miles, which commeth neere the ſubſtance of our pretence, and ſhall be farre more eaſie then the premiſes.

First come to the place where the ground ſerueth you beſt to place your Canon, there make a marke for your firſt ſtation, then ſend ſome ſpare bodie a conuenient diſtance off, that waie, which you finde the ground fitteſt, ſo it be not directlie backwards or forwards, to ſet you vp another mark for your ſecond ſtation. Which done, you ſhall with your ſtandard ſtaffe after the firſt manner of the laſt Chapter, take the angle of poſition betweene your ſecond ſtation, and ſome one pinacle or other notable marke of the Caſtle or forte, thereat locke your ſtaffe by helpe of the centre ſcrewe at that angle that you may carry him ſafely without ſagging, vnto your ſecond ſtation. This done, you ſhall goe to the other ſtation, meaſuring the diſtance betweene with a wyre chaine, more pikes, or how you will, but moſt exactly, be ſure. Thoſe perches or pike lengths you ſhal reckon on that leg of your ſtandard ſtaffe, which you before directed to your ſtation, by the euen parts and numbers theron ſet: and to the end of that reckoning place, the running bolt or ſight of that legge, and there fixe him by help of his ſcrew pin. Then place your ſtaffe again at this ſecond ſtation on his ſupporters, ſo that this fixed ſight pin be directly ouer the marke of the ſecond ſtation, and the centre end directed towards the firſt ſtation, which your partner ſhall ſupport, and lifte vp the other legge withall, putting to and fro the running ſight of that legge by your commandement

Familiar Staffe.

ment and direction, vntill that your selfe doe finde the foresayd marke, hole or pinacle, directly euen with the knops or tops of both the running sight pins. And then looke what number of partes you finde there shewed in that further legge by the second sight pin: so many paces, perches, or pike lengths is that marke or pinacle fro your first station.

P. the first station where you woulde place your Canon K. or rather N. the second station. T. a marke in the tower to bee battered. R. and S the two supporters, A B C the standard staffe placed on his supporters, his centre A at the first station placed ouer the marke P, his one legge A C directed to the second station at K his other leg A B to the marke T in the tower. T A C, the angle of position taken at the station P. Admit the distance betweene the stations P and K, to be fortie perches or pike lengths that number fortie I seeke in the legge A C, it endeth at N, there I set the running sight N fixed, which at this second station K, I place iust ouer K, directing the centre A to P (the angle of position B *A* C vnstirred) M. the running

The vse of the

ning sight of the other legge, set directly betweene the pinacle T, and the sight N: so shall the number of partes comprehended betweene A and M on the leg A C, shew you the exact distance from the first station A, vnto the pinacle, in such measure as you measured between your stations P and K.

Many things might be noted, which in halfe a daies practise will be easier found, then though I shoulde write many nights.

1 First, the greater distance that your two stations are in sunder, except they be vnreasonable much further then the marke: the more certaine is your working.

2 The sharper that the angle is made betweene the legges of the Staffe, as for example, F A K in this second figure or the blunter, as G A K: The greater distance of the stations is required: otherwise the point of Crossing either at F or G, will hardly appeare.

For these causes you must vse as much discretion in pointing your second station as the ground will allow you. The best is, when your marke maketh an equicrurall triangle with your stations or neare about, as D A N and this requireth least distance betweene your stations. Wherefore if you may chuse your ground, point your second station so, that you neuer open the legges of your staffe much more or lesse then the right angle, though you make your stations the shorter: a litle lesse is

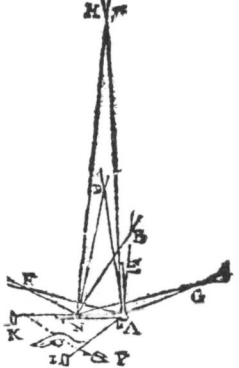

alwaies best, but neuer much bigger then a right angle if you can possible chuse: if so, then make your stations the longer.

4 If you meane to get an exact distance, you had néede (yea by your best position) let the distance betweene your stations be no lesse then one eight part of the distance to your
marke

Familiar Staffe.   35

marke as neare as you can imagin. Otherwise you shall hardly get the true point of crossing, as you may perceiue at H: or at the least let it be neuer lesse then one twelfe parte: either of which are easie to be had in a small ground, and will serue for a great distance: for fortie pearches is but a reasonable distance for two stations, and are to be found commonly in a small close, and yet that is the eight part of a myle of a thousand pace, and it is the twelfth part of your common longest vsuall myles.

5 Other small conceites are to be noted as if the distance of the stations in the last example were so small, that in account on the legge A C of your staffe it falleth out so neare the handle, that the running sight N might not be brought thereto: then may you set your sight N at double that number, and when you haue done, take halfe the number found on the legge A B betweene A and the sight M for youre distance desired, or if you set him at thrise or foure times, and in the end take a third part or quarter of that you finde, all commeth to one passe. Yea if the distance be very short, you may reckon euery tenne but for one, and then if you reiect one cyphare from euery number your staffe is readie figured to your contentment.

6 One thing I had almost forgotten: for as in rekoning the distance of your stations on the legge A C you shall euer be rather led to encrease your scale, as in the fift note I shewed, especially, for short distances or longitudes, as they call them: So contrary wise for long distances, when as the mark desired is a myle or two from you, except your stations bee made much the larger, you shall be driuen to place halfe, or a third, or a quarter of your stationary distance betwéene A and N: and then take the double, treble, or quadruple of that you finde betweene A and M, for your desire, as in steade of the fortie perches betwéene P and K to haue set twentie or ten euen partes betwéene A and N. Yea, or if your stations had béene fortie seuen: (here because fortie seuen is not easily parted into a quarter because of the fraction) you might first

F 2   worke

worke by a quarter of the euen number, that is to say of fortie, and then worke by the seuen by it selfe, and adde the two productes, as if by ten you found an hundreth perches, you must take foure hundreth for those: and then working by the odde seuen, you shall get seuentie more to bee added to foure hundreth, so is your whole distance foure hundreth and seuentie.

Also you must prouide to haue two bright markes flickering ouer both your stations till you haue done, that you bee assured iustly to take your leuell.

Chap. 16.
Of the Geometricall ground, and familar proofe thereof, wheron the vvhole working by this familliar Staffe dependeth.

AS I haue thought it very requisite to shew the grounde platte of these precepts in hand, so it were in vaine for mee here to iterate any expositions or diffinitions of vsuall termes Geometricall, since they are now a dayes very common, and are of Euclide and diuerse other recited, and of me in my Mathematicall Iewell 1. booke sufficiently declared. Vnderstand therefore that if in any triangle you drawe a line parallell to any one side, the same line shal cut of a triangle like, & equiangled to the former.   For example within the triangle A B C

let there bee drawen the line D E parallell to C B I say that y͛ triangle A D E is equiangled to the triangle A B C, that is to say, the angle

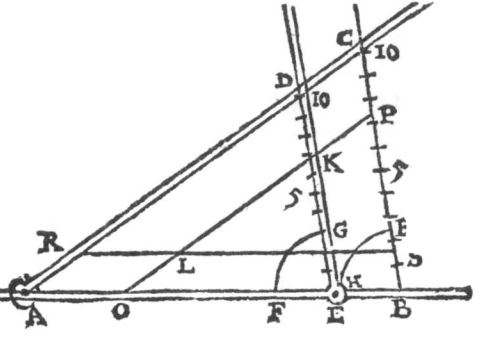

E is

Familiar Staffe.   37

E is equall to the match angle B, and the angle D to the angle C, and the angle A of the one, to the angle A of the other, for the angle A is common to both, & also the triangle A D E, shall be like to the triangle A B C, that is to say, what proportion the sides of A D E beare one vnto another, the like proportion doe the sides of A B C beare one to another. For Familiar proofe of the first, open your compasses at auenture, and with one scantlet on the centres B and E, describe two seuerall arches cutting the seuerall sides, encluding the same angles B and E, as you see F G and H I, and those two arches you shall finde by your compasse to be iust equall the one to the other, and therefore of necessity the angles F and B subtended by those arches are equall: in like manner may you proue the angle D to be equal vnto the angle C, and being the angle A common to both, therefore the triangles are equiangled, if you desire more artifyciall proofe, repaire to Euclide lib. 1. propos. 29. and lib. 6. propos. 5. 6. and 7. Likewise for Familiar proofe of the second, deuide any one side of the triangle A B C into some nomber of euen parts, admitte the side B C into ten equall parts, then deuide the match side of the other triangle into the like number of parts, viz. D E into ten equall parts also, then measure with your compasse how many of those parts of B C are contayned in the side A B of the triangle A B C, and you shall be sure to finde the like nūber of the parts of D E in the match side A E of the triangle A D E, and as many parts of B C as are in A C, so many parts of D E are in A D: therefore are they saide to be like and proportionall. This Familiar proofe I haue thought beste for my Familiar Staffe: for artificial proofe seeke Euclide, lib. 6. propo. 4. 5. 6. and 7. Euen so in the figure of the last Chapter at the second station K, the leg A B of your standerd Staffe doth cut the maine triangle T K P or rather

F 3   T N A,

The vſe of the

T N A by the line M A parallell to the ſyde T A, making within the iuriſdiction of the Staffe, the triangle N M A equiangled, and like to N T A. Therefore being that N A is appointed ſo many euen partes as K P contayneth euen perches, it followeth of neceſſity, yͤ what euen parts you finde betweene M A, there muſt alſo be ſo many euen perches in T A, which is the longi-tude or diſtance deſi-red. Note that the letter A wanteth in this figure at the ſta-tion K.

Note alſo that if in the former figure you drawe the line O P parallell to A C, and R S parallell to A B, they ſhall make a litle triangle in the bely of A B C, viz. K L M equiangled, and like to A B, yea and euery triangle that you can picke out in the ſame figure is equiangled, and like to A B C, and therefore each of them to other. Thus haue I opened you a great breach, which if you throughly entre and poſſeſſe, you may be able alwaies to atchieue theſe maiſtres without booke at all times of neede.

Chap.

Chap. 17.

How to performe the fifteenth Chapter with more facillity by meanes of the second manner of taking the angle of position mentioned in the 14. chap.

LEt no man be amazed at this working for it is all one in substance with the 15. Chapter, but oppositely performed, which in diuerse partes of the Mathematickes produceth the same troth as the direct way doth. For many times Oppositorum eadem est ratio, if I shall familiarly shewe you the difference it may be compared vnto a cupple of men going together towardes one place, the one with his face forwardes, the other backwardes, the matter is no more difficult, therefore marke it well. Let P and K be your two stations, & T the tower whose distance you seeke from your first station P. You shall therefore at P with your standerd staffe after the second manner of the 14. chap. take the angle of Position betweene T and your second station K, or rather L directly ouer it, as in this figure you see C A B, and locke fast your staffe at that angle of position, for it is the opposite angle of T A L, and equall vnto it by Eucl. lib. 1. propos. 15. as in the 14. chap. was shewed. Then repayre vnto your second station K, measuring the exact distaunce betwéene

Of longit. or distance.

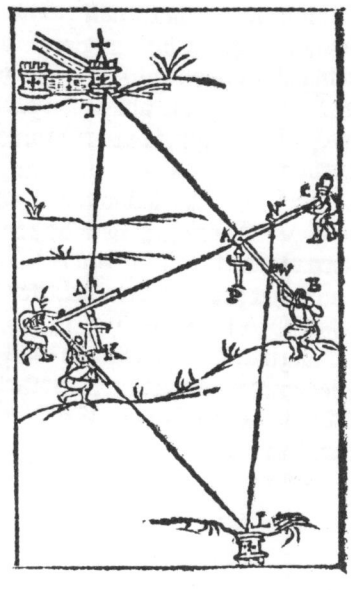

The vſe of the

your ſtations with ſome wyer cheyne or moryce pykes, and according to that meaſure place the running ſight of that leg of the ſtaffe, which at P was directed to K, viz. at L. This ſight L there fixed, you ſhal now place directly ouer your ſecond ſtation, by reſting it on your dagger or ſome ſticke ſtuck into the ground, and let your partner (applying the centre of the ſtaffe to his eye) direct the ſame legge againe to the firſt ſtation P by helpe of the ſight pegges, as here you ſee. And then your ſelfe ſhall moue too and fro the running ſight of the other legge vntill you may place it ſo that by it you may ſee the tower T, euen with the ſight before fixed at L, viz. at O, by the viſuall line O L T. To conclude then, I ſay, that the euen part of the ſtaffe cut by the ſight O, ſhall ſhew you the exact diſtance of the tower T from your firſt ſtation P. The certainty hereof may well be proued by the laſt chap. for the angle L A O is all one with the angle N A M becauſe the ſtaff is faſt lockt to that angle, and therefore equall to T A L being but the oppoſite angle of N A M the angle A L O being alſo oppoſite to T L A, is therefore equall vnto it by Euclyde lib. 1. propoſ. 15. and when two angles of any tryangle are equall vnto two angles of another tryangle, the thirde angles can not chuſe but bee equall by Eucl. lib. 1. propoſ. 26. then ſeeing the tryangle L A O is equiangled to L A T therefore of neceſſitie they muſt be like, and their ſides proportionall by Eucl. lib 6. Prop. 4. and A T beareth that proportion to A L as O A doth to A L of the tryangle O A L, as if within the tryangle A T L you did ſet the tryangle A L O you ſhould eaſily conceaue. Or if you turne the figure vpſide downe, then ſhall the working ſeeme all one with that of the fifteenth Chapter, the Angle A M N being equall to A L O.

Chap.

Familliar Staffe. 27

Chap. 18.

In case that a Castle or fort were digged out of some rocke, or scituate in some valley betweene two hilles, admitting but some narrow comming vnto it, and therefore enforcing your two stations to be one directly behind another. Hovv yet by this familiar Staffe you shall attaine the distance of the same from you.

First come as neare the wall as you can without danger, so that it be not ouer neare, and take your running staffe set at his square angle, and both the winges placed on the hanging side. Apply the extreame of his leuel side to your eye, carrying him as leuell as you can, and there looke where a-boute you may place the vpper wing to see the toppe of the wall, but yet at some euen number of fiues or tennes be sure because of after reckoning. The vpper wing so placed to remaine fixed, then let your partner and you together settle your staffe a litle further or neare (his hanging side kept plumbe) till you may from the extreame of the leuell side, see the highest part of the wall or tower euen with the said fixed wing. Marke that station on the ground. Then goe backe a good quantitie, and there chuse another station, doing in the very same manner with the other wing, placing him also fixed at some euen ten or fiue. Then measure exactly the distance betwéene those two stations: For looke so many times as the euen partes or distance betweene the two winges, are contained amongst the euen partes from the centre of the staffe to the vpper wing: so many times is the measurement betweene your two stations contained in the whole length betwéene your furthest station and the wall.

Of longitude or distance.

G As

42   The vſe of the

As for example, admitte that the diſtance betwéene the ſtations M and N were twentie pearches, and that at the inner ſtation N you ſet the one wing, to wit L at two hundreth perches from D the centre of the ſtaffe, and there fixed. And at M the outer ſtation, you ſet the other wing, that is to ſay H at an hundreth and fiftie partes: Your ſtations M and N ſo choſen, that at N you ſee the toppe of the towre, to wit P, by the viſuall line E L P, and at M, by the viſuall line E H P. The number of euen partes (by taking an hundreth and fiftie out of two hundreth) I finde fiftie, betweene the winges L and H. Now I conclude becauſe L H that is to ſay fiftie, is foure times contained in L D: two hundreth therefore foure times the diſtance of M N that is, foure times twentie perches is the true diſtance from M the outermoſt ſtation vnto the towre, to wit, the length of M R.

    Another way is thus, get the altitude of the towre top aboue the leuell of your eye, that is to ſay of P S by the ninth or tenth Chapter, which had, reckon the ſame on the hanging ſide of your running ſtaffe, and thereto ſet the wing fixed, admit at H. Then chooſe you a ſtation as farre from the towre as you thinke good, admit at N, there direct your ſtaffe againſt the towre as before, kéeping the ſtation N directly vnder

Familliar Staffe. 43

der your eye, and your eye no higher leuell then S, then moue the wing of the leuell side, that is I to and fro, till you may by both the sightes I and H, see the top of the towre that is to say P. Then looke how many euen partes are betweene T and D, for so many paces is the distance from your station at N to the towre.

Note that if there be no wall or towre high inough for the exact working of this Chapter farr inough off: you may make a shift to it, by two markes of the wall in latitude, that is to say in the length of the wall: for that the working by the altitude and latitude is al one, as in the end of the tenth Chapter was noted, so you can direct your staffe square thereto: but if you happen not to direct him square, you misse so much of your distance to the wall as maketh vp the square: as for example. If T P were a wall, and that aiming at the square, you should direct your staffe to the marke W thinking the square to be there, I say you shall in stead of the distance E W, get the distance E S, such is the certaintie of your crosse staffe working, which you can not auoid because of necessity, as in the second Chapter was said In transuersis visui credere oportet.

Chap. 19.

How you shall peerforme the 15. Chapter, vvhere the distance of the fort or castle is verie far off, and the ground being vneuen vvith hils, dales and rockes, hath no one leuell plaine sufficient to make two stations for so great a distance.

IF some valley happen betwéene your first station, and that place where you woulde willingly, according to the third   *Of longitude or distance.*

44   The vſe of the

third note of the 15 Chapter. Plant your ſeconde ſtation, as here you ſee in this figure deſcribed, a déep bottome betwéen the ſtations P and K. In this caſe you ſhal by a third ſtation made on the plaine, whereon the firſt ſtation is, viz. at Q and help of the firſt, get the diſtance of the fartheſt ſtation K, beyond the ſayd vallie, in ſuch ſorte, as in the 15. or 17. Chapter is taught, or as in the next Chap. ſhalbe ſhewed. And ſo ſhall you haue the diſtance of your 2. long ſtatiōs

P and K, as exactly as though they were in one plaine to bée meaſured with a line. And by this meanes, though the valley be neuer ſo déepe betwéene your two maine ſtations P and K, it ſhall not hinder you, but that you may at them procéde to get the moſt exact diſtance to the forte or tower, in all reſpects as in the 15. or 17. Chap. was done, euen as though the ground had bene all one plaine.

Chap. 20.

If in a night you haue ſecretlie gotten with your armie neere a-
mie forte, and that you woulde with more ſpeede then in the

15. or

Familiar Staffe. 45

15. or 17. Chapter is shewed, know the distance whether you are neere inough to plant your ordinance for batterie.

NOw commeth in the readie vse of the running staffe, for which cause I so named him. And in this Chapter resteth the sweete vse of this instrument, therefore marke it well. This matter resteth wholie vppon the diligent furtherance of many helpers both of men, and horses also, if you will, according to the prouerbe, Fit cito per multas præda petita manus. There is no more cunning in this, then was in the fifteene Chapter: rather lesse. But that you must haue diuerse to execute it: that is to saie, a couple (with horses if you will) to speed them to the second station, there by helpe of the runninge staffe, to take the angle of position betwéene the first station, and some turret of the forte, whiles your selfe and your partner are as busie at the first station with your standarde staffe, taking the position angle between the same turret and the second station, which will most speedely bee done by the seconde meanes of the 14. Chapter, and sie vpon restes or supporters nowe we are in hast. The whiles also let two other speedilie running with a wier chainze, get the exact measure betweene the two stations: or if the forte be far off, or the ground vnleuell, it will not be amisse to get the distance between those two stations by a thirde station, in manner of the last Chapter. Which distance either way most ex-

Of long or dista[nce]

G 3   actly

actly gotten, let him that keepeth the first station reckon it on one of the legges of the same standard staffe, it is no matter, which (for now we be in haste, wee must not be too curious in precepts) and to the end of that reckoning, let him place the running bolt or sight, fixed with helpe of his screwe. And if by this time his partener bee not come from his second station, then let him take a nagge and gallop to meet him, taking speciall care that either of them locke their staues fast at the angles of position before taken, least they should start with this hastie carriage. Now then when they are met, if he that came from the second station, doe but clap the centre of the running staffe on the sight pin of the standard before fixed, guiding it so that his one side cut the centre of the standard, extending his other side crosse the standards other legge. Looke at that crossing what partes you finde, for the partes on the standard are the distance from your first station to the forte, and the partes of the running staffe, are the distance of the second station, to the same fort desired, which is so much more then the standard could doe of himselfe, besides the speed, yet for all your hast, the notes of the 15. Chapter must be regarded. And thus much farther, that you enlarge the leuell side of your running staffe with the ioynt peece of the hanging side taken off, if cause so require.

For example: admit C A B to be the angle of the position, taken by the standard at the first station, and D E A the angle of position, taken by the running staffe at the second station, and that on the one leg A B of the standard, the point E, reckoned from A, expresse the distance between the two stations mesured: there shall the

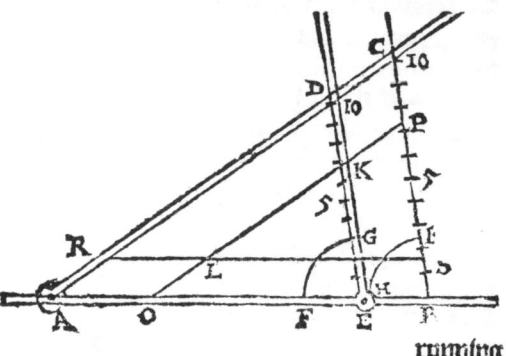

running

Familiar Staffe. 47

running bolte or sight pin be set, and therein the centre E, of the running staffe A E D his one side, lying euen with the standards legge A B, and cutting his centre A, so doth the other E D cut off in the legge A C at D the distance of the tower from the first station. And in the point D of the side E D of the running staffe cut off by the standards leg A C, sheweth the distance of the tower from the second station.

Chap. 21.

If you shall see two fortes of the enemies within viewe, and woulde know how farre they are in sonder, and whether ther may be passage for an army betweene them without danger of shotte from those fortes. Or to get the length and breadth of any Forte a far off, thereby to gather of what receite the same forte is. Or to get the width of a riuer fronting any fort, keeping your selfe a farre off without daunger of shotte.

FIrst you shall by the fifteenth, seuenteenth, nineteenth, or Of latitude last Chapter, get the seuerall distances of the same fortes from your station appointed, seeke the one distance in the one legge, the other distaunce in the other legge of your standerd Staffe, and to each sett the running sighte pins fired, then pitching againe at the same station, take with your standerd Staffe the angle of position betweene those fortes by the fourteenth Chapter, and there wreste him faste to keepe that angle sure. Lastly place the centre of the running Staffe on one of the said fixed sight pinnes, applying also one of his sides vnto the other sighte pin, and the number of that side of the running Staffe, there touching that other sight pin is the distaunce betwene the two forts desired. Or you might apply the ioincte piece or any parte of the running Staffe betweene the sight pinnes, but that it is not so easy to rekon as from the centre it is.

In like manner may you get the distaunce betweene two trees or two towers in one sorte, and thereby get the length

48   The vſe of the

or bredth of the ſame ſorte, and be able by that meanes to deſ-
crie of what receit the ſame is. Alſo the bredth of any riuer
by taking two markes, as trées, buſhes, or buildings on each
ſide of the riuer one. But you are by this neuer the neare,
to know the diſtance betweene two ſhippes on the ſea, as o-
ther writers and thoſe no ſmall ones, do teach you by a like
precepte arithmeticall, except thoſe ſhippes lie at ancor that
they ſtirre not, yet in the twenty foure Chapter following, I
ſhall ſatisfie you for that.

For example, admitte the diſtaunce of the one ſorte founde
to bee fiue
hundred paſe
from your ſta-
tion, the o-
ther ſire hun-
dred paſe: the
firſt viz. fiue
hundred I re-
kon on the
one legge of
the ſtanderd,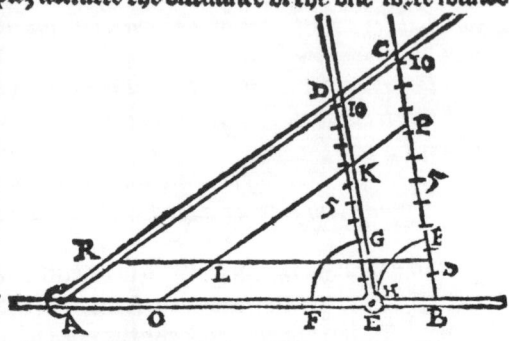
there I ſette the ſighte pinne viz. at E, the ſecond viz. ſire hun-
dreth I reckon on the other legge of the ſtanderd, there I ſet
the other ſighte pin viz. at D. That doen, I take the angle
of poſition betweene both thoſe ſortes, or rather ſome two
principall towers of them, which admitte to be the angle C
A B, there I lock the ſtanderd Staffe, then clapping the centre of the running Staffe on the ſight pin before fixed at E, I
directe his owne ſide vnto the other ſight pin fixed at D, and
the quantity thereof betwéene D and E, ſheweth me the diſtaunce betweene the two ſortes or towers deſired. In like
manner the diſtaunce betweene two trées, is to be had or betweene any two marks whatſoeuer.

This is another maner of working for latitudes then that
which the Croſſe Staffe yieldeth : the manner whereof I
noted vnto you in the end of the tenth Chapter.

Chap.

Familiar Staffe. 49

Chap. 22.

If in a fort or hauen on the sea coast, or abroad on the plaines on the sea banckes, you shall see anie ships a farre off sayling towardes you, or any armie approching by lande. Howe you shall alwaies be prouided in such speciall places, that in a moment almost and with small helpe, you shall know how farre they are from you, and by that meanes speedily finde when they shall be commen within the random or pointe blancke of your shot.

IN such dangerous times when any assault of an army by land or of ships by sea is expected against any fort or hauen, you shall appoint in the same hauen or fort two certaine towers farthest distant, in stead of your two stations. Their exact distaunce you shall alwayes haue in a reddinesse gotten by the last chap. Let one of these towers be called the first station the other the second. In this first tower you shall haue alwayes your standerd readie placed by some kind of steadfast supporter, his one legge directly on the other tower or station, not to be easely wrested aside from that position, the other leg next the enemy at liberty. In the second tower let the running staffe be alwaies readily placed in like manner, his one side stedfastly on the first station, his other side next the enemie at liberty. Your instruments thus exactly placed, alwayes readie at your two stations, and the certaine distaunce betweene them foreknowen, your worke is at any time almost performed ere you begin. Let your selfe at the first station, and some practizd body at the second, guide the loose legs of your instrumentes directly still vpon the mayne toppe of some principall shippe, or formost Auncient of the army, and let a third man within a little after he shall perceaue both your eyes layde to your instruments with a trumpet or horne, giue a short blast: and immediatly vpon the instant of the blast, let both of you locke fast your instruments at those angles, which angles had

Of longit. or distance.

H   and

and taken both at one instant of the blast giuen. Let each of the obseruers come from their towers and bring their instrumentes to méete together as spéedily as they can, and clapping one vppon another in such manner, as in the 20. chap. was done, you shall there immediatly descry how farre that shippe or army was off at the blast giuing.

But because it may be long ere two can come out of the toppes of towers to méete: besides that, because it is harde bringing such an instrument downe stayers, or letting it downe with a rope for feare of sagging or slipping of the angles taken. Therefore to remedie this, let him that obserueth in the first tower, haue a spare running staffe, besides the standerd staffe wherewith he obserueth: And let him that giueth the blast, bee placed néere about the middest betwéene these two station towers. And when the blast is giuen let him in the second tower that obserueth with the running staff, looke what degrée of the graduation on the leuell side of his staffe, the graduators Ayer doth shew, and report it with a loude voyce to this middle man: And let him report it againe to the obseruer in the first tower or station, who hauing the sayde spare running staffe shall immediatly set it by the same degrée and minute reported vnto the very same angle that the running staffe at the second tower is at, and therewith worke in all respects as before euen as though the other running staff had beene brought to him.

But if any man will obiect, what if there bee not two towers in the hauen, then I say it is likely there is but one Church, and that it is a small matter in such daungerous times to make scaffoldes with bordes on housen toppes for such good purposes.

Also in the very like spéedie manner, you may worke with a cople of stations readie pitched on the sea bankes: withall you must euer haue regarde to the notes of the 15. chap.

Chap.

Familiar Staffe.

Chap. 23.

If an army on the land, or a Nauie on the sea, shall be as farre off as you may ken making towards your Fort. To know by helpe of this Familiar staffe: How fast their gate is, and in what time they shall according to that gate, come within reach of your shot.

This matter very little differeth from the last chap. For your two staues being readie planted in their towers or stations as there is shewed. First take the distance of the army or Nauy, as farre as you can possibly ken them, at a blast or sound giuen by the middleman, whose charge shall further be for this purpose to tourne vp a mynute glasse or two, so sone as he hath giuen his blast, and to kéepe them running. Then after a quarter of an howre or 15. mynutes or more or lesse time as you will, and as the occasion requireth, prepare your selues to take the distance of that nauy or army at another short blast giuen in the very same maner as before. Now hauing gotten the distance of the shippe or army this seconde time: & also vpō report of the middle man how many mynutes were passed betweene his two blasses giuen, you may easely know by the difference of the two distances taken, how much they be come nearer at this second blast then they were at the first, for so much do they saile or come on in so many mynutes, and according to that rate you may easily cast how much they sayle in an howre, and consequently howe many howres or mynutes it will be, if holding on their way they approch you, and in how many mynutes they will be within the randome or point blancke of any great piece, and rather then to be deceaued least the winde or state of the seas should alter, you may make more tryalls ere they come neere that.

But in stead of mynute glasses, which as they saye, they vse at the sea, I should like better of a clocke or watch that should turne the hand quite round about the dyall euery houre to shew the exact mynute I haue séene such readie made.

Of motion in longit.

The vſe of the

Chap. 24.

If ſtanding on the ſea bancks, you ſee your ſhip of war at the ſea, purſuing another ſhip of the enimy, to know how far by help of this Familliar Staffe, & how much the one getteth of the other in ſayling, whether he be likely to ouertake the enimy, and in what time.

Of motion both in lōgitude and latitude.

This matter differeth litle from the working of the laſt Chapter, onely it asketh more helpe. For you had néed to haue at the firſt towre or ſtation thrée ſtanderd ſtaues, and foure obseruors, and at the second tower or ſtation two running ſtaues, and two obseruors all ready pitched and placed, as in the two & twenty Chapter was taught, in ſuch ſort, that one paire, that is to ſay, one ſtanderd Staffe, and one running Staffe doe wholy attend vpon your ſhip, the other payr vppon the enemies ſhippe, and the odde standard Staffe to attend them both, by helpe of two obseruors, to kéepe alwayes the angle of poſition betweene the two ſhippes, and al of them to locke faſt their angles at the inſtante of a blaſt giuen, and then by the two and twenty Chapter to get the diſtaunce of ether ſhippe from you, and by the one and twenty, and helpe of the angle of poſition, the diſtaunce of either ſhip from other at the time of the blaſte. Then after a quarter of an howre or certaine minutes compleate, they muſt al to worke agayn by warning of a ſecond blaſte, as well for the diſtances of eche ſhippe from you, as of one of them from the other. Then comparing the differences of ech kinde together, you haue by the one (that is to ſay) by their ſeuerall diſtances from your ſtation in that number of minutes, as were betwéene the two blaſtes geuen, the certainty of each of their gates, in ſo much time by the laſt Chapter, and by the two ſeuerall diſtances of ech ſhippe from other at both times, you may by taking the bigger out of the leſſer, know how much the one ſhippe hath

eſher

Familiar Staffe. 53

ether gotten or losse of the other, and according to that rate of time, cast how soone the one shall ouertake the other, if the wind and state of the seas alter not, or they alter not theyr course. For which cause you shall make some more trialles in the very like manner as oft as you shall thinke good.

Chap. 25.

Your selfe planted on the toppe of an high rocke, clift, or tower, by the sea side. To knowe by this Familliar Staffe howe deepe the leuell of the water is vnder you.

IT is most common vppon the sea coaste to haue highe rocky cliftes. If therefore you woulde from the top of anye such knowe how deepe directly vnder you, that the leuell of the water is, you shall espy some marke in the bottome of the sea bancks or next clifte, euen with the edge of the water the nearest vnto you that you can espy. But if there be none neare inough, then let some ship be lette purposely at ancor nere vnto you, the distaunce of which ship, you shal get by two pretty stationes, such as the toppe of the rocke may yielde by the fifteenth or seuenteenth Chapter. Or if the rocke or clifte bee so stiepe and sharpe, that it yield you not station inough on the toppe, see then if you can discend to some lower part of the rocke for a second station, or any way by any rule in this booke, helpe your selfe to the distaunce of the shippe from your station on the hill toppe. Which had, take your standerd staffe, and let him at a true square angle by helpe of your running Staffe: there locke or wreste him faste that he start not from his square, Then place him so, that his one legge bee leuell or parallell to the horizon, the other perpendiculer sette by a plumme line, his centre ouer your station. Your standerd so placed, then your partner and you directing the leuell legge towards the shippe or marke: you shall between you moue the running sighte pinnes of both legges to and fro in such sort, till by them you see the saide marke or shippe where

Of profundity.

H 3   the

The vſe of the

ſhe lieth at ancor, there wreſte the ſight pinnes faſt. Then ſette the centre of the running Staffe on the ſight pin of the leuell legge, directing his one ſide to the middle or centre of the other ſight pin, his other ſide euen with the centre of the ſtanderd. There locke the running Staffe faſt at that angle, and marke what number of the running Staffe is there cut off by the centre of the ſighte pin. For if that be equall to the diſtaunce of the ſhippe or marke before gotten, then doth that ſighte pinne ſhew your deſire, if otherwiſe then moue the running Staffe faſt lockt at the ſaid angle to and fro, the centre of the ſtanderde by helpe of the ſighte pinne whereon hee rideth, vntill there be cutte off in the ſaide ſide of the runninge Staffe, the exacte diſtaunce of the ſaid ſhippe or marke before found. And then ſhal the pointe of cutting ſhewe you on the ſtanderd the very height of your eie aboue the leuell of the water deſired.

This matter would haue an example. Admitte firſt on the cliſte A, whoſe depth to the leuell of the water is A B, that you finde the diſtaunce from your eie on the toppe of the cliſte, to the ſhip ſett from you at ancker one hundreth twenty perches, viz. I C, and hauing ſette vp your ſtanderd Staffe, viz. O V I faſte lockte to his ſquare angle, his one legge viz. I V, perpendiculer, his other viz. O V leuell, his centre V ouer the

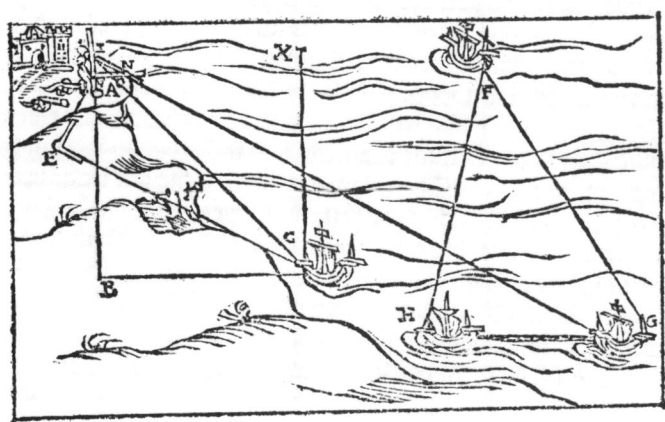

Familiar Staffe.   55

ſtation, viz. A as before is ſaide, then by remouing his ſighte pinnes to ſuch two places, viz to I and O, that by your eye you finde the ſhippe directly in one line by them both, viz. in the line I O C, and that the ſighte pinne in the perpendiculer legge, viz. I doe ſhewe 100. partes, the other in the leuell leg three hundreth partes. Now take your running Staffe and ſette the centre thereof on the ſighte punne O, and directe his owne ſide vnto I, and his other ſide euen with the centre V of the ſtanderd, and there locke him faſt at that angle. And if now you finde on the running Staffe betweene O and I the iuſt number of the diſtaunce I C, viz one hundreth twentye, then doth I V ſhew the exacte number of perches of I B deſired. But in very deede you ſhall finde betweene I and O, three hundreth ſixteene; or thereabouts, which is more then one hundreth twenty. Therefore you ſhall moue the ſight pin O, together with the running Staffe inwards towards V, vntill ſuch time as you finde but iuſte one hundreth and twenty betweene I and O, and then ſhall you haue but thirty eight iuſt betweene I and V, your deſire. For thereby I conclude that the depth from I to B is thyrty eyghtperches.

This Chapter ſeemeth more painefull at the firſt ſhewe, then it wil be when it is rightly vnderſtoode, and that you haue made experience thereof by an example or two the beſt is, it is needefull to bee doen but once for al, in anye one place.

Chap. 26.

If a Gunner keepe a block houſe, or haue a piece or two of Ordinance planted on the toppe of ſome very high clifte by the ſea ſide. How by this Familliar Staffe he him ſelfe without any helpe at all, ſhall moſt eaſily and ſpeedily in a moment get the diſtaunce of any ſhip at the ſea, making towards him or paſſing along, and
                                                                that

The vse of the

that most exacte, for so far as the randome of any great piece will extend.

Of longitude or distance frō an high.

TRuely this Chapter on such cliftes and places on the sea coste, which are very high aboue the water (as many such there are) is the most necessarie and ready for a gunne of all the rest. But if it be not high aboue the sea, it is least worth: And the higher, so much it is the better for this purpose, and also for the Gunner, as it should seeme vnto me by Tartaglia who saith in his first booke, second Colloquy, that a peece either mounted or embased from the leuel of the Horizon shooteth with more force and further, then it wil, being shot leuell. The chiefe substance of our practise in this, consisteth in the exact finding of the leuell of the water vnder your station on the clifte toppe: For which cause I promised the last Chapter. Whereas others haue left it very absurdly to be done with a line and plummet of lead let down from the clift toppe: as though euery clift or any clift almost would admit such palpable facillitie.

    Well, to the purpose. In such a set place you shall haue alwaies your station ready prepared, with the depth of the leuell of the water knowne by helpe of the last Chapter: and there your standard setled at his square angle, and alwaies ready pitched in maner of the last Chapter, on some such rest that he may speedely be turned which way soeuer any shippe commeth, keeping him selfe still plumbe and leuell, the sight pinne of the perpendiculer side, fixed fast with his scrue at the euen part, answering the hight of your station aboue the water, your standard staffe thus alwaies set in a readinesse, your worke is done in manner ere you beginne. For so soone as you espye any shippe making neare to your coste, direct the extreame of the leuell legge of the standard to the shippe, and clap on the centre of your running staffe, to ride on the fixed sight pinne of the hanging legge, and there lifting vp or downe the one side of the running staffe close by the said leuell legge

Familliar Staffe.   57

legge directed to the shippe, vntill by the sight pegges of that side of the running staffe you see the shippe. Then shall the euen part of that side of the running staffe, cut by the leuell legge of the standerd, be the distaunce of the shippe desired, al which by this figure you may perceaue.

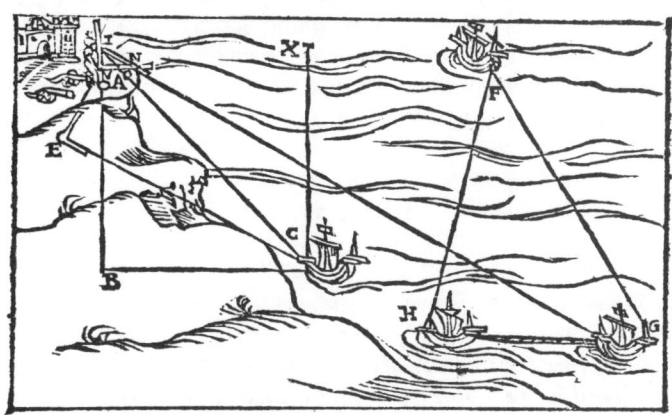

A the station on the clifte N VI, the standerd Staffe alwaies ready placed thereon his leuel legge V N directed to the shippe G approching, I the sight of the standerds hanging leg, whereon the centre of the running Staffe is placed, whose one side I N, directed close by the standerds legge V N to the bottome of the shippe, G is cutte off by V N at N. Thereby I conclude that the euen parts betweene I and N sheweth the distaunce from your eie to the ship viz. I G desired.

Note that the other side of the running Staffe is to no vse in this practise. And one thing more I thought good to warne you off, that in placing of your sight pin I, you haue a care of the rising or falling of the leuell of the water by reason of the tides. For which cause, you had neede to haue some speciall markes purposely sette in the bottome of the next clifte or sea

I   bancks

58 The vſe of the

banckes, by which you may at any time gather how much the water is riſen, or fallen, and according to that riſing or falling you muſt fixe the ſighte pin I higher or lower when you begin your worke.

    Note that this Chapter and the laſt alſo, mighte as well haue bene performed in manner of the twelfth Chapter, by turning the extreame of the hanging ſide of the ſtaffe downewards if the place doe admitte, and happily that courſe maye ſeeme the more familiar, and propper way for profundities although both come to one paſſe.

Chap. 27.

If a man were priſoner with the enemie, how being in the top of a tower on the leads, or out of his priſon window, hee might by this Familiar Staffe know the depth to the ground, to ſee if he were able with anie deuice to let himſelfe downe without danger.

Of profundity.
    FOr this purpoſe, yea, or almoſt anie precept of altitudes, an inſtrument of ten inches long, is as good as one of ten foot, and ſuch a one a man might priuily haue conueyed vnto him, and eaſily wilde it in a chamber. You ſhall out at a windowe by the eighteenth Chapter, or from the toppe of the leads, by the fifteenth or ſeuenteeth chapter, get the direct diſtance from you of ſome tower next yours, which ſuppoſe were X C, whoſe baſe C you may ſee out at your windowe, viz. IX, then place your inſtrument plum and leuell towardes that tower X C, in ſuch ſort as was done in the laſt Chapter. But contrariwiſe to the laſt Chapter, place that diſtance I X, gotten on the leuell ſide at O fixed. Then putting vp or downe the ſight pin of the hanging ſide, viz. I, till by it and the ſight pin O, you

Familliar Staffe.

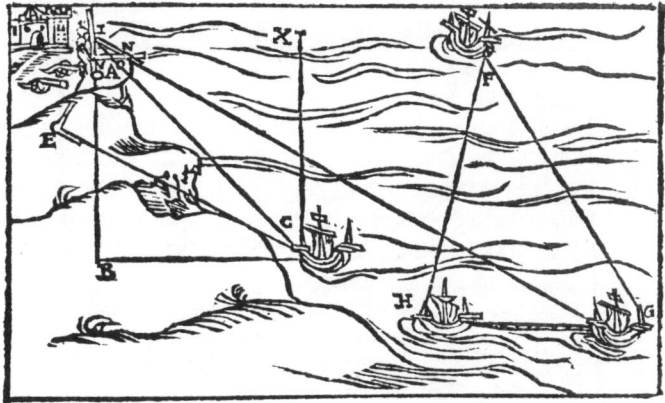

O, you may sée the base of that tower, viz. C, by the visuall line I O C. There I saie, shall the sight pin I shew you the deapth of the tower XC, viz. from x the leuell of your eie, vnto C the base, and so high from the ground you may presume that your selfe are in that roome where you are prisoner. Let your selfe downe and you can, sauing your necke péece.

Happily this Chapter were more easily performed by turning the hanging side of the Staffe downwards, according to the note in the end of the last Chapter, because the instrument is small, and may easily rest in the one end.

Chap. 28.

If a fort or tower stand vpon an high hil how by this Familiar staffe, to know the ioynt and heights both of the hill and tower.

TO helpe vs for want of a figure, you shall admit that Y were the top of the hill, and base of the tower, and P Y the the tower standing thereon. Which granted. The first thing you shall doe, get the distante from your standing, viz. N, vnto the point S leuell with your eie, directly vnder the base of the tower,

Of altitude and distāce of altitude.

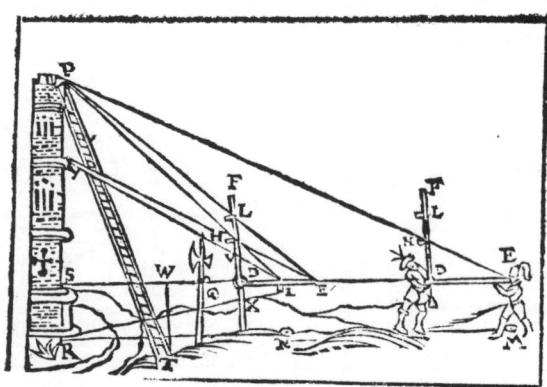

tower, by the eighteenth Chapter. But if the hil and tower be farre off, then if by the fifteenth or seuenteenth Chapter you get the distance I Y, and by discretion take somwhat lesse then that for the distance I S, a little error in that longitude wil not hurt vs for the altitude. This distance I S anie waie gotten, you shall reckon the same on the leuell side of your running Staffe E D directed to the hill, and there set the one wing, viz. L, and from it direct your vsuall line to the towers base, viz. to Y, and set the other wing on the hanging side directly betwéen I and Y at V. I saie the number at V shewed by that wing, is the height of the hill desired, viz. of Y S, and if from I you direct your vsuall line againe to the top of the tower on the hill, viz. to P. Then shal the wing on the hanging side cut at H, where agayne you shal haue the whole height both of tower and hill viz. of P S, and the distance betwéene H and V shèweth the height of the tower P Y seuerall.

Chap. 29.

If being at the sea, you would cast ancor as neere some fort or harbour, as you might be free from reach of their shot,.

Familiar Staffe. 61

shot, how by this Familiar Staffe you shall exactlie get the distance thereof, of the distance of anie other sh p from your ship, being both fleeting at once vppon the wilde sea.

This is the most curious matter which we had to deale withall, and as sleightly dealt in by other writers, whose whole deuice hetherto, hath béene to looke from the top of the ship wherein they are, vnto the bottome of the other ship, performing it after the same reason that in my 26. Chapter is shewed. For they appoint the knowen length of the mast of their ship, in stead of the knowen height of the clift in that 26 Chapter. And to make the matter seeme of more account, they call this a distance gotten at one station. A goodly thing: as who saith that the square angle, which they imagine at the bottome of the maste, and the length of the mast knowen, is not in all respects a second station, as it were readie, made to theyr hand. Let me alwaies haue a reentangled triangle to deale with, and I will neuer aske you but one such station, for all that I haue before written. But now looke yet neerer vnto the matter, and though the ground of this their working is vnfallible, yet see to how small purpose, and in manner absurd, theyr precept is in this. For if (as in the fourth note of the foureteenth Chapter I shewed) the second ship bee distant aboue ten or twelue masts length, their working is of small account, euen on plaine ground: but then consider the tottering of the ship on the water, they shall be forced for want of steddy taking of the angle euen *totalliter errare*. For which causes I cannot content my selfe to deliuer this matter so sleightlie, as that I thinke it to anie purpose to be done in anie ship, but rather by helpe of two shippes, or in a calme one ship and help of the cockbote. Let both those ships bee grapled or linked together with a strong rope or cable, and let sailers and sternes men, keep the ship with side winds, and labouring the ruthers or stearnes at the full length of the rope, facing the fort, har-

I 3   borow

borowe, oz other ship, whose distance you seeke, and at the two ends of those two ships farthest distant, let the standard staffe

Heere should stand the picture that is in the
27. Chapter.

in the one, and the running staffe in the other, bee placed with two obseruors to each staffe, to take the angle of position, after the second manner of the fourteenth Chapter, betweene the sayde forte and his fellow obseruors, in the other ship sodenly at a blast giuen, as was done in the 22. Chapter, there locking fast their instruments, and hasting by haling in the ships to bring them together, they shall descrie the distance to the forte desired, in such sort as in the 20, or 22. Chapter was done, or without haling the shippes together by reporting the angle taken one to another, and help of a spare running staffe, as in the end of the said 22. Chapter is shewed. For the length of the cable and both ships knowen, yeldeth the distance betweene your two stations.

And for the more spéedy and sure taking of those angles, let each payre of obseruors hang a verie white cloth vnder the centre of their instruments. In this figure H and G are the two ships linked together with a cable rope, and F is a ship at sea, whose distance they séeke.

Note that if a man be so straight driuen, that he hath no helpe but one bare ship, then let him doe it in the verie lyke manner, placing his obseruors at both endes of the ship, as far

of

Familiar Staffe. 63

of as may be, which shall performe his desire: much more certayne then from the top of the mast. And if anie man woulde doe it from the mast, then let him take the angle downe to the bottome of the ship with a sea Astrolabe, and after set his running staffe to the same angle, & his standard staffe to a square angle, and then performe it in manner of the fiue and twenteth Chapter.

Chap. 30.

How by this familiar staffe to carry the leuel of one place to any other, necessarie for such as shal vndermine a fort, to know alwaies how deepe they are: or for such as would trie whether waters may bee brought from one place to fortifie another.

This matter is verie easie, requiring more diligence then skill. For if you set the leuell side of your staffe leuel with the Horizon, by helpe of a plumme line, and hauing a couple of assistants to carie two poles before you, with bright marks sliding vp and downe on them. Then all the matter consisteth in the diligent noting of the rising or falling of the ground, by helpe of those moueable marks carried from place to place on those poles, vntill you come to your waies end: and then shal you compare them together, whether is the greater of all the risings or all the fallings, each added seuerally together into a grosse summe. For if they be equall, then the leuel at the end is equall to the leuell at the beginning: if vnequall, the difference sheweth the difference of the rising or sinking accordingly.

And if you should carrie the leuel many miles, to the intent to lead water in a conduite, then for the fall thereof, you must haue also a regarde of the roundnesse of the massie globe of the earth and water. Of which I shall at some other time more at large signifie.

Chap.

The vse of the

Chap. 31.
How by helpe of this familiar Staffe, you shall carrie a mine vnder the ground, and sett barrells of gunpowder directlye vnder any towre or chiefe place of any castle or forte.

This is no rare matter neither harde to performe. First gette the exacte distaunce to the same towre from the place where you meane to begin your mine, by the fifteenth, seuenteenth, or twenty Chapter. Then sincke your selfe into the ground at the mouth of the mine, as much vnder as you think meete, and there by two plumlines from aboue pitched directly vppon the middle of the towre, sette the leuell side of your running Staffe vnder ground directly vppon the towre, by which direction you shall cary your mine streight till you come almost there. And if you be hindred that you can not digge streight in any place, by meanes of some rocke make a square, retourne by direction of your Staffe, till you are past the same rocke, and you must of necessity when you come almost there, make two or three square retournes out of the way, and still bring your selfe to your direct way againe, to the end to stay the force of the powder for issuing directy backe out of the ouen or caue closed vp, thereby hindring his exploite. So keeping your note booke certainly of your goings aside out of the direct way, you may by the like square angles retourne againe iust as much vnto the same, and then proceed direct againe. And also keeping notes how you goe either shallow or deep, in manner of the last Chapter, you shal at the last both knowe when you are gone so far as the distance first taken aboue ground amounted to: and also how deepe or shallow you are from the bottome of the towre to be blowen vp.

Chap. 32.
How a Captaine may by this familiar staffe, sett in platte or mappe any prouince of the enemies couutrey.

This matter is all ready handled of many writers to be don by the circle, quadrante, or crosse Staffe, but specially by the

Familiar Staffe.

the circle, and is in our Englishe tongue published by diuerse namely, Master Digges, William Boorne, Master Lucar, and others, vnto whom I therefore refer you for the circumstaunce, because they shall not bee idle the whiles. Onely thus much I thoughte good to note vnto you, that look what angles of position they take by theyr cyrcle, the same may you and your partner more speedily take by the second manner of the foreteenth Chapter, and your running Staffe, and note the degrées of euery angle or position in a booke or tables as they doe. And when you haue those angles noted, and would reduce them into mappe, if you can not otherwise more readily doe it, then take paines in such manner as they teach, by helpe of that small cyrcle, which they call, and fitly a protractor, in the mean while till I may haue a wished time to write of this my Instrument more at large. Then will I teache you to lay a side your detractor, and shewe you more knacks then I dare at this time to name or make promise of for my owne quiet sake, leaste I should be by some importunacyes interrupted from the seconde part of my Iewell, which I woulde moste gladly bring forth with as conuenient spéede as may bee.

K   An

An Appendix of the author, touching some alteration of this Familliar Staffe.

Auing on the foure and twentieth of Ianuary last, presented the coppy of this smal treatise in written hand, together with ye Staffe, framed in such manner, as in the third and fourth Chapter is shewed. It pleased his Honor within two daies after, to see triall of some of the conclusions in Greenewich parke. Upon the good liking whereof, his H. shewed the same to diuerse, both Noble, Honorable, and worshipfull, in the Courte: and withall deliuered me againe the written coppy, as a thing which he thought more meete to be made publike, for the benefite of his countrey, then to be reserued to his owne priuate vse. But I well noted that the generall opinion of the greater number of best iudgement, was, to haue him made in two seuerall staues, in such manner as in the ende of the thirde or fourth Chapter I noted, thereby to haue them both of a more conuenient tractable quantity. Which I more willingly consente vnto, because I finde some wante of a thirde slippinge rule in the staderd Staffe, like vnto the graduator of the running Staffe, to keep him sure and stedfast in working: which after the first fabrication may not be wel conueied.

But now if we shall like of this alteration of our Staffe: there are also many things of the third and fourth Chapter to be quite altered.

The first is, that you haue two seuerall channells in the hollownesse of your Staffe, the vndermost for the graduator, and the bolte leading his aper to and fro : the vppermoste for the running bolts that must carry your sighte pinnes to and fro, which bolts shall also now haue your wings or broad sightes

Familiar Staffe.   67

sights of brasse (appointed for the running Staffe) so annexed vnto them, that they may folde or shut into the Staffe, when you carry it.

Secondly the graduators of these seuerall staues shall haue no maner of degrees or graduation vppon them except you list.

Thirdlie there shall be no ioincte piece like vnto G F to bee taken off and on, but both legges shall bee whole and sounde quite through.

Fourthly, wheras the chiefest cause of the ioincte piece G F was in respecte to be taken off, when the running Staffe was to be vsed for the leueling or mounting of a piece of ordinance, in manner of the sixth Chapter. We shall supply that vse (though both legges of the Staffe be whole) by setting the gunners points on the leuell side D E, by helpe of a quadrant, described on the pointe G, as in the fourth Chapter you did sette them on the graduator G H, by helpe of a quadrant described on the pointe D, the Staffe standing still at his square angle. Or els (which I think better) you may set the Staffe vnto the angle of forty fiue degrees, and there fasten him, and then on N describe a quadrante, and thereby sette on, either the twelfth gunners points, or the 90. degrees on the side D E as before, and so shall a line drawen from G to the middle pointe of the twelfth, bee perpendiculer to D E: for which cause you shall now number them, from that middle pointe, but vnto six on either side, because the one half of these twelue pointes, shal nowe serue for mounting a piece vnto the sixth

H 2   pointe,

pointe, which as Tartaglia setteth downe is the furthest randome, the other halfe serueth for embasing your piece vnto the sixt point, vnder the leuell or pointe blancke, in all respects to be vsed as in the sixt Chapter is shewed, sauing that now the hanging side D G F must be turned vpwards, which there was turned downwards, and the angle F D E sette at the angle of forty fiue degrees, which there was set at the square angle.

Note that if you liste, you may sette the pointes of mounting, beyond the sixe point, as far as the length of the side D E will extende.

Touching the proportion and the timber fit for this Staffe.

I Thought it not amisse to note, that making your Instrument into two seueral Staues, the fittest proportion that I like, is, that each legge of your Staffe be fiue or sixe feete in length, one inche broad, and one inche and an halfe thicke to receaue the channells.

The one channell for the graduator to be shutte into, off, halfe inche wide, the other for the bolts of the running sights, to slip to and fro in: of one quarter of an inche wide, and the three cheekes enclosing these two chanels, to bee eche one quarter inche thicke: so is your whole inch and halfe thicknes bestowed.

Now for him that will bestowe cost, let each of these legs be made of fiue rule pieces, that is to say, three, to serue for the said three cheekes, which must be of peare tree or plum tree, as well for tough strength, as to receaue the grauinge, the other two pieces serue to fill vp or backe the channelles which must be of Cedar, or some such very light and stronge wood, to the end your Staffe be not ouer heauy. Also for a Captaine it may be fashioned, out to a sharpe pointe aboue the ioinre(not much vnlike the number of a paire of comnon

fire=

Familiar Staffe.   69

fire-tongs) to the end, to receaue thereon the hed of a pike or leading Staffe.

Or if anie man will bestowe somewhat lesse cost, then let each leg bee made but of three rule peeces, and haue but one channell, the two out sides of peare trée, the middlemost of ceader as before. But then he must haue some forecast, that the boltes for his running sight pins, and the bolt that leadeth the graduators aper to and fro, may all runne and worke theyr wils in that one channell, which I know may be done.

Lastly, for the common sort that will bestow little cost, let each leg be made but of one péece, keeping the former breadth and thicknesse, and let the channell or channels bee cut out of the whole woode, but then make him not ouer long, least hee sagge. For assure your selfe, the more péeces he is made of, the stronger he is from sagging, and the surer from warping. And howsoeuer hee bee made, let him at the least be brought to an eight square, or rounded, for the more pleasant handling.

To conclude, anie thing here set downe for hast without figures, that may not be well conceiued or gathered by the premises, my selfe will be most readie to giue directions for, vpon anie small request. I shall easily bee heard of about maister Treasurers lodging in the Court, or at Swallowfield by Reading, where I dwell. There dwelleth a verie artificiall workeman in Hosier lane, called Iohn Reade, who can further you, whose helpe I haue vsed about one or two of these staues.

Touching Marine causes.

Amongst many that tooke view of my Familiar Staffe at the Court, the right noble and worshipfull Knight, Sir George Carew, Knight Marshall of England, and chiefe Captaine of the Ile of Wight, tooke no small notice of his necessarie parts, and required me instantly to apply him as well for the vse of the sea, as I had don for the land: which amongst a number of other Astronomicall and Geometricall conclusi-

K 3   ons

ons, I meant to haue reserued for my second edition. But yet to satisfie his Worthinesse and others of like affection: let it be knowen, that for taking the altitude of the Sunne, Moone, or Starre. If you put a short pegge into the centre of your Staffe, and rest the same peg on the very vtter corner of your eie, and then open the legs of your staffe, vntill (with your eie so rested on the same peg, and your other eie closed) you sée by the one leg or his sight pin, the brim or parting of the sea from the element: by the other leg, the Sunne, Moone, or Starre desired, there shall the graduators aper shew you, among the degrées or graduation of the Staffe, the altitude desired.

Note that your Staffe being shut almost close, you may by a thred fastned to the Aper, pull the Staffe open at pleasure.

Then for taking the distance of two Starres or Planets, which as I take it, is a second matter vsed at sea. You cannot possibly (as I gather) do it better by the Crosse staffe it self, then you and your partner may doe it, after the second manner of the fourteenth Chapter before.

I coulde also shewe you meanes, by applying the extreame of the leuell side to your eie, and pulling the graduators aper towards you, how you might take the altitude or distance of starres by sinking of the hanging side, by lyttle and little from the square angle, in maner, Crosse staffe wise: but these things, with a number other, for hast I let passe, presuming that the ingenious wil easily finde them out of theselues.

To make a finall end, if for these Marine causes you make my staffe no longer, then their common Crosse staues are, I see no cause, but it is as readie, and may be handled with as much facilitie as their Crosse staffe. And if yet there remaine anie doubt or defect, I make no doubt, but vppon aduertisements thereof to supplie it.

Farewell my Familliar Staffe, commend me to mine old acquaintance *Richard Stockwel*, Gunner of the Ile of *Gernsie*, if thou méete him, tell him thou art prouided to satisfie him of al our communication had at *Greyes* this last Summer, and let him lodge thee in the stored shrine of all his pretie conceits.

FINIS.

TA
579
B39
1590a

OCT 1 3 1971